Differential Equation over Banach Algebra

Aleks Kleyn

Aleks_Kleyn@MailAPS.org
http://AleksKleyn.dyndns-home.com:4080/
http://sites.google.com/site/AleksKleyn/
http://arxiv.org/a/kleyn_a_1
http://AleksKleyn.blogspot.com/

ABSTRACT. In the book, I considered differential equations of order 1 over Banach D-algebra: differential equation solved with respect to the derivative; exact differential equation; linear homogeneous equation. In noncommutative Banach algebra, initial value problem for linear homogeneous equation has infinitely many solutions.

Copyright © 2018 Aleks Kleyn

All rights reserved.

CreateSpace Independent Publishing Platform

ISBN: 1983521639

ISBN-13: 978-1983521638

Translated from Russian
Дифференциальное уравнение над банаховой алгеброй
Александр Клейн

Contents

Chapter 1. Preface 4

Chapter 2. Preliminary Definitions 5
  2.1. Universal Algebra 5
  2.2. Representation of Universal Algebra 6
  2.3. Module over Ring 8
  2.4. Algebra over Commutative Ring 10
  2.5. Derivative of Map of Banach Algebra 13
  2.6. Quaternion Algebra 15
  2.7. Direct Sum of D-modules 16
  2.8. Direct Sum of Banach D-Modules 22

Chapter 3. Differential Equation Solved with Respect to the Derivative . 27
  3.1. Differential Equation $\dfrac{dy}{dx} = 1 \otimes x + x \otimes 1$ 27
  3.2. Differential Equation $\dfrac{dy}{dx} = 3x \otimes x$ 34
  3.3. Before Going Any Further 42
  3.4. Forms of representation of differential equations 42
  3.5. Condition of Integrability 44

Chapter 4. Differential Equation of First Order 48
  4.1. Differential Equation with Separated Variables 48
  4.2. Exact Differential Equation 51
  4.3. Linear Homogeneous Equation 55

Appendix A. Summary of Statements 59
  A.1. Table of Derivatives 59
  A.2. Table of Integrals 63

Appendix. References 65

Appendix. Index 67

Appendix. Special Symbols and Notations 68

CHAPTER 1

Preface

When I began to study noncommutative algebra, I could write a linear map only together with its argument. Tensor representation of linear map allowed me to simplify notation and expanded the field of my research. Now I am ready to study differential equations over Banach algebra.

Since the product is noncommutative, the set of equations which I can consider is more limited than in the commutative case. However the set of differential equations considered in this book is source of extremely interesting theory.

Derivative of map in Banach spaces has different representations: if we consider a map f of D-algebra A represented using coordinates with respect to basis of D-module A, then derivative is represented by Jacoby matrix of the map f; if we consider coordinateless representation of the map f, then derivative is represented by $A \otimes A$- number. So I decided to write this paper such way that somebody can use it regardless of the presentation, and, in examples, I want to consider solving of differential equation in different representations.

I dedicated the chapter 3 to differential equation solved with respect to the derivative. To make the picture complete, I consider different forms of representation of differential equation and compare the results. Examples in the text give ability to see how calculations work in noncommutative algebra. At the same time, these examples are source of new ideas and they are an integral part of the book.

The statement that problem with initial condition for linear homogeneous equation has infinitely many solutions was the most interesting and unexpected statement.

CHAPTER 2

Preliminary Definitions

This chapter contains definitions and theorems which are necessary for an understanding of the text of this book. So the reader may read the statements from this chapter in process of reading the main text of the book.

2.1. Universal Algebra

DEFINITION 2.1.1. *For any sets*[2.1] *A, B,* **Cartesian power** B^A *is the set of maps*
$$f : A \to B$$
□

DEFINITION 2.1.2. *For any $n \geq 0$, a map*[2.2]
$$\omega : A^n \to A$$
*is called n-***ary operation on set** *A or just* **operation on set** *A. For any a_1, ..., $a_n \in A$, we use either notation $\omega(a_1, ..., a_n)$, $a_1...a_n\omega$ to denote image of map ω.*
□

DEFINITION 2.1.3. *An* **operator domain** *is the set of operators*[2.3] Ω *with a map*
$$a : \Omega \to N$$
If $\omega \in \Omega$, then $a(\omega)$ is called the **arity** *of operator ω. If $a(\omega) = n$, then operator ω is called n-ary. We use notation*
$$\Omega(n) = \{\omega \in \Omega : a(\omega) = n\}$$
for the set of n-ary operators.
□

DEFINITION 2.1.4. *Let A be a set. Let Ω be an operator domain.*[2.4] *The family of maps*
$$\Omega(n) \to A^{A^n} \quad n \in N$$
is called Ω-algebra structure on A. The set A with Ω-algebra structure is called **Ω-algebra** A_Ω *or* **universal algebra**.
□

[2.1] I follow the definition from the example (iv) on the page [12]-5.
[2.2] Definition 2.1.2 follows the definition in the example (vi) on the page page [12]-13.
[2.3] I follow the definition (1), page [12]-48.
[2.4] I follow the definition (2), page [12]-48.

DEFINITION 2.1.5. *Let A, B be Ω-algebras and $\omega \in \Omega(n)$. The map*[2.5]
$$f : A \to B$$
is compatible with operation ω, *if, for all* $a_1, ..., a_n \in A$,
(2.1.1) $$f(a_1)...f(a_n)\omega = f(a_1...a_n\omega)$$
The map f is called **homomorphism** *from Ω-algebra A to Ω-algebra B, if f is compatible with each $\omega \in \Omega$.* □

DEFINITION 2.1.6. *A homomorphism in which source and target are the same algebra is called* **endomorphism**. *We use notation* $\mathrm{End}(\Omega; A)$ *for the set of endomorphisms of Ω-algebra A.* □

CONVENTION 2.1.7. *Element of Ω-algebra A is called A-**number**. For instance, complex number is also called C-number, and quaternion is called H-number.* □

2.2. Representation of Universal Algebra

DEFINITION 2.2.1. *Let the set A_2 be Ω_2-algebra. Let the set of transformations $\mathrm{End}(\Omega_2, A_2)$ be Ω_1-algebra. The homomorphism*
$$f : A_1 \to \mathrm{End}(\Omega_2; A_2)$$
of Ω_1-algebra A_1 into Ω_1-algebra $\mathrm{End}(\Omega_2, A_2)$ is called **representation of Ω_1-algebra A_1** *or A_1-**representation** in Ω_2-algebra A_2.* □

We also use notation
$$f : A_1 \dashrightarrow A_2$$
to denote the representation of Ω_1-algebra A_1 in Ω_2-algebra A_2.

DEFINITION 2.2.2. *Let the map*
$$f : A_1 \to \mathrm{End}(\Omega_2; A_2)$$
be an isomorphism of the Ω_1-algebra A_1 into $\mathrm{End}(\Omega_2, A_2)$. Then the representation
$$f : A_1 \dashrightarrow A_2$$
of the Ω_1-algebra A_1 is called **effective**.[2.6] □

DEFINITION 2.2.3. *Let*
$$f : A_1 \dashrightarrow A_2$$
be representation of Ω_1-algebra A_1 in Ω_2-algebra A_2 and
$$g : B_1 \dashrightarrow B_2$$

[2.5] I follow the definition on page [12]-49.

[2.6] See similar definition of effective representation of group in [15], page 16, [18], page 111, [13], page 51 (Cohn calls such representation faithful).

2.2. Representation of Universal Algebra

be representation of Ω_1-algebra B_1 in Ω_2-algebra B_2. For $i = 1, 2$, let the map
$$r_i : A_i \to B_i$$
be homomorphism of Ω_i-algebra. The matrix of maps $(r_1 \ r_2)$ such, that

(2.2.1) $$r_2 \circ f(a) = g(r_1(a)) \circ r_2$$

is called **morphism of representations from** f **into** g. We also say that **morphism of representations of Ω_1-algebra in Ω_2-algebra** is defined. □

REMARK 2.2.4. *We may consider a pair of maps r_1, r_2 as map*
$$F : A_1 \cup A_2 \to B_1 \cup B_2$$
such that
$$F(A_1) = B_1 \qquad F(A_2) = B_2$$
Therefore, hereinafter the matrix of maps $(r_1 \ r_2)$ also is called map. □

DEFINITION 2.2.5. *If representation f and g coincide, then morphism of representations $(r_1 \ r_2)$ is called* **morphism of representation** f. □

DEFINITION 2.2.6. *Let*
$$f : A_1 \relbar\joinrel\twoheadrightarrow A_2$$
be representation of Ω_1-algebra A_1 in Ω_2-algebra A_2 and
$$g : A_1 \relbar\joinrel\twoheadrightarrow B_2$$
be representation of Ω_1-algebra A_1 in Ω_2-algebra B_2. Let
$$\left(\mathrm{id} : A_1 \to A_1 \quad r_2 : A_2 \to B_2 \right)$$
be morphism of representations. In this case we identify morphism $(\mathrm{id} \ r_2)$ of representations of Ω_1-algebra and corresponding homomorphism r_2 of Ω_2-algebra and the homomorphism r_2 is called **reduced morphism of representations**. *We will use diagram*

(2.2.2)
$$\begin{array}{ccc} A_2 & \xrightarrow{r_2} & B_2 \\ {\scriptstyle f(a)}\uparrow\downarrow & & \downarrow{\scriptstyle g(a)} \\ A_2 & \xrightarrow{r_2} & B_2 \\ \nearrow{\scriptstyle f} & {\scriptstyle g}\nearrow & \\ A_1 & & \end{array}$$

to represent reduced morphism r_2 of representations of Ω_1-algebra. From diagram it follows

(2.2.3) $$r_2 \circ f(a) = g(a) \circ r_2$$

We also use diagram

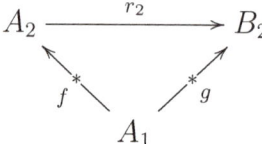

instead of diagram (2.2.2). □

2.3. Module over Ring

DEFINITION 2.3.1. *Effective representation of commutative ring D in an Abelian group V*

$$f : D \rightarrow\!\!\!* V \quad f(d) : v \rightarrow dv \tag{2.3.1}$$

is called **module over ring** D *or* **D-module**. *V-number is called* **vector**. □

THEOREM 2.3.2. *Following conditions hold for D-module:*
2.3.2.1: **associative law**

$$(pq)v = p(qv) \tag{2.3.2}$$

2.3.2.2: **distributive law**

$$p(v + w) = pv + pw \tag{2.3.3}$$

$$(p + q)v = pv + qv \tag{2.3.4}$$

2.3.2.3: **unitarity law**

$$1v = v \tag{2.3.5}$$

for any $m, n \in Z$, $a, b \in D$, $v, w \in V$.

PROOF. The theorem follows from the theorem [10]-4.1.3. □

DEFINITION 2.3.3. *Let $\overline{\overline{e}}$ be the basis of D-module V and vector $\overline{v} \in V$ has expansion*

$$\overline{v} = v^*{}_* e = v^i e_i$$

with respect to the basis $\overline{\overline{e}}$. D-numbers v^i are called **coordinates** *of vector $\overline{v}$ with respect to the basis $\overline{\overline{e}}$.* □

DEFINITION 2.3.4. *Let A_1, A_2 be D-modules. Reduced morphism of representations*

$$f : A_1 \rightarrow A_2$$

of D-module A_1 into D-module A_2 is called **linear map** *of D-module A_1 into D-module A_2. Let us denote $\mathcal{L}(D; A_1 \rightarrow A_2)$ set of linear maps of D-module A_1 into D-module A_2.* □

2.3. Module over Ring

THEOREM 2.3.5. *Linear map*
$$f : A_1 \to A_2$$
of D-module A_1 into D-module A_2 satisfies to equalities[2.7]

(2.3.6) $$f \circ (a + b) = f \circ a + f \circ b$$
(2.3.7) $$f \circ (pa) = p(f \circ a)$$
$$a, b \in A_1 \quad p \in D$$

PROOF. The theorem follows from the theorem [10]-4.2.2. □

THEOREM 2.3.6. *Let*
$$\bar{\bar{e}}_1 = (e_{1 \cdot i}, i \in \boldsymbol{I})$$
be a basis of D-module A_1. Let
$$\bar{\bar{e}}_2 = (e_{2 \cdot j}, j \in \boldsymbol{J})$$
be a basis of D-module A_2. Then linear map
$$\overline{f} : A_1 \to A_2$$
has presentation

(2.3.8) $$b = a^*{}_* f$$

relative to selected bases. Here

- *a is coordinate matrix of A_1-number $\bar{a}$ relative the basis $\bar{\bar{e}}_1$*

(2.3.9) $$\bar{a} = a^*{}_* e_1$$

- *b is coordinate matrix of vector*

(2.3.10) $$\bar{b} = \overline{f} \circ \bar{a}$$

  relative the basis $\bar{\bar{e}}_2$

(2.3.11) $$\bar{b} = b^*{}_* e_2$$

- *f is coordinate matrix of set of vectors $(\overline{f} \circ e_{1 \cdot i}, i \in \boldsymbol{I})$ relative the basis $\bar{\bar{e}}_2$. The matrix f is called **matrix of linear map** relative bases $\bar{\bar{e}}_1$ and $\bar{\bar{e}}_2$.*

PROOF. The theorem follows from the theorem [9]-5.4.3. □

DEFINITION 2.3.7. *Let D be the commutative ring. Let A_1, ..., A_n, S be D-modules. We call map*
$$f : A_1 \times ... \times A_n \to S$$
polylinear map *of modules A_1, ..., A_n into module S, if*
$$f \circ (a_1, ..., a_i + b_i, ..., a_n) = f \circ (a_1, ..., a_i, ..., a_n) + f \circ (a_1, ..., b_i, ..., a_n)$$
$$f \circ (a_1, ..., pa_i, ..., a_n) = pf \circ (a_1, ..., a_i, ..., a_n)$$

[2.7]In some books (for instance, [1], p. 119) the theorem 2.3.5 is considered as a definition.

$$1 \leq i \leq n \quad a_i, b_i \in A_i \quad p \in D$$

□

DEFINITION 2.3.8. *Let A_1, ..., A_n be free modules over commutative ring D.*[2.8] *Consider category $\mathcal{A}_1$ whose objects are polylinear maps*

$$f : A_1 \times ... \times A_n \to S_1 \quad g : A_1 \times ... \times A_n \to S_2$$

where S_1, S_2 are modules over ring D, We define morphism

$$f \to g$$

to be linear map

$$h : S_1 \to S_2$$

making following diagram commutative

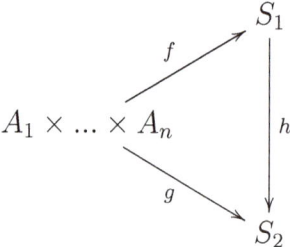

Universal object $A_1 \otimes ... \otimes A_n$ of category $\mathcal{A}_1$ is called **tensor product** *of modules A_1, ..., A_n.*

□

2.4. Algebra over Commutative Ring

DEFINITION 2.4.1. *Let D be commutative ring. D-module A is called* **algebra over ring** D *or* D**-algebra**, *if we defined product*[2.9] *in A*

(2.4.1) $$v\,w = C \circ (v, w)$$

where C is bilinear map

$$C : A \times A \to A$$

If A is free D-module, then A is called **free algebra over ring** D.

□

THEOREM 2.4.2. *The multiplication in the algebra A_1 is distributive over addition.*

PROOF. The theorem follows from the theorem [10]-5.1.2.

□

The multiplication in algebra can be neither commutative nor associative. Following definitions are based on definitions given in [19], page 13.

[2.8] I give definition of tensor product of D-modules following to definition in [1], p. 601 - 603.

[2.9] I follow the definition given in [19], page 1, [11], page 4. The statement which is true for any D-module, is true also for D-algebra.

2.4. Algebra over Commutative Ring

DEFINITION 2.4.3. *The* **commutator**
$$[a, b] = ab - ba$$
measures commutativity in D-algebra A. D-algebra A is called **commutative**, *if*
$$[a, b] = 0$$
□

DEFINITION 2.4.4. *The* **associator**
(2.4.2) $$(a, b, c) = (ab)c - a(bc)$$
measures associativity in D-algebra A. D-algebra A is called **associative**, *if*
$$(a, b, c) = 0$$
□

DEFINITION 2.4.5. *The set*[2.10]
$$N(A) = \{a \in A : \forall b, c \in A, (a, b, c) = (b, a, c) = (b, c, a) = 0\}$$
is called the **nucleus of an** *D-algebra A.* □

DEFINITION 2.4.6. *The set*[2.11]
$$Z(A) = \{a \in A : a \in N(A), \forall b \in A, ab = ba\}$$
is called the **center of an** *D-algebra A_1.* □

CONVENTION 2.4.7. *Let A be free algebra with finite or countable basis. Considering expansion of element of algebra A relative basis $\overline{\overline{e}}$ we use the same root letter to denote this element and its coordinates. In expression a^2, it is not clear whether this is component of expansion of element a relative basis, or this is operation $a^2 = aa$. To make text clearer we use separate color for index of element of algebra. For instance,*
$$a = a^i e_i$$
□

CONVENTION 2.4.8. *Let $\overline{\overline{e}}$ be the basis of free algebra A over ring D. If algebra A has unit, then we assume that e_0 is the unit of algebra A.* □

THEOREM 2.4.9. *Let $\overline{\overline{e}}$ be the basis of free algebra A_1 over ring D. Let*
$$a = a^i e_i \quad b = b^i e_i \quad a, b \in A$$
We can get the product of a, b according to rule
(2.4.3) $$(ab)^k = C_{ij}^k a^i b^j$$
where C_{ij}^k are **structural constants** *of algebra A_1 over ring D. The product of basis vectors in the algebra A_1 is defined according to rule*
(2.4.4) $$e_i e_j = C_{ij}^k e_k$$

[2.10] The definition is based on the similar definition in [19], p. 13
[2.11] The definition is based on the similar definition in [19], page 14

PROOF. The theorem follows from the theorem [10]-5.1.9. □

DEFINITION 2.4.10. *Let A_1 and A_2 be algebras over commutative ring D. The linear map of the D-module A_1 into the D-module A_2 is called* **linear map** *of D-algebra A_1 into D-algebra A_2.*

Let us denote $\mathcal{L}(D; A_1 \to A_2)$ set of linear maps of D-algebra A_1 into D-algebra A_2. □

DEFINITION 2.4.11. *Let $A_1, ..., A_n, S$ be D-algebras. Polylinear map*
$$f : A_1 \times ... \times A_n \to S$$
of D-modules $A_1, ..., A_n$ into D-module S is called **polylinear map** *of D-algebras $A_1, ..., A_n$ into D-algebra S. Let us denote $\mathcal{L}(D; A_1 \times ... \times A_n \to S)$ set of polylinear maps of D-algebras $A_1, ..., A_n$ into D-algebra S. Let us denote $\mathcal{L}(D; A^n \to S)$ set of n-linear maps of D-algebra A_1 ($A_1 = ... = A_n = A_1$) into D-algebra S.* □

THEOREM 2.4.12. *Let A be D-algebra. Let product in algebra $A \otimes A$ be defined according to rule*
$$(p_0 \otimes p_1) \circ (q_0 \otimes q_1) = (p_0 q_0) \otimes (q_1 p_1)$$
A representation
$$(2.4.5) \qquad h : A \otimes A \twoheadrightarrow \mathcal{L}(D; A \to A) \quad h(p) : f \to p \circ f$$
of D-algebra $A \otimes A$ in module $\mathcal{L}(D; A \to A)$ defined by the equality
$$(a \otimes b) \circ f = afb \quad a, b \in A \quad f \in \mathcal{L}(D; A \to A)$$
allows us to identify tensor $d \in A \otimes A$ and linear map $d \circ \delta \in \mathcal{L}(D; A \to A)$ where $\delta \in \mathcal{L}(D; A \to A)$ is identity map. Linear map $(a \otimes b) \circ \delta$ has form
$$(2.4.6) \qquad (a \otimes b) \circ c = acb$$

PROOF. The theorem follows from the theorem [10]-6.3.4. □

CONVENTION 2.4.13. *I assume sum over index i in expression like*
$$a_{i.0} x a_{i.1}$$
□

THEOREM 2.4.14. *Let $\overline{\overline{F}}$ be basis of left $B \otimes B$-module $\mathcal{L}(D; B \to B)$. Let*
$$G : A \to B$$
be linear map of maximal rank. The set
$$\overline{\overline{F}} \circ G = \{F_k \circ G : F_k \in \overline{\overline{F}}\}$$
generates the set of maps
$$(2.4.7) \qquad \{g \in \mathcal{L}(D; A \to B) : \ker G \subseteq \ker g\}$$

PROOF. The theorem follows from the theorem 2.4.14. □

THEOREM 2.4.15. *Let $\overline{\overline{e}}_1$ be basis of the finite dimensional D-module A_1. Let $\overline{\overline{e}}_2$ be basis of the finite dimensional associative D-algebra A_2. Let $f \in \mathcal{L}(D; A_1 \to A_2)$. Let C_{kl}^p be structural constants of algebra A_2. Let $\overline{\overline{F}}$ be the basis of left $A_2 \otimes A_2$-module $\mathcal{L}(D; A_2 \to A_2)$ and $F_{k \cdot i}^{\ j}$ be coordinates of map F_k with respect to basis $\overline{\overline{e}}_2$. Let*

$$G : A \to B$$

be linear map of maximal rank such that $\ker G \subseteq \ker f$ and G_i^j be coordinates of map G with respect to bases $\overline{\overline{e}}_1$ and $\overline{\overline{e}}_2$. Coordinates f_l^k of the map f and its standard components $f^{k \cdot ij}$ are connected by the equation

(2.4.8) $$f_l^k = f^{k \cdot ij} F_{k \cdot r}^{\ m} G_l^r C_{im}^p C_{pj}^k$$

PROOF. The theorem follows from the theorem [10]-6.4.5. □

2.5. Derivative of Map of Banach Algebra

DEFINITION 2.5.1. *Normed D-algebra A_1 is called **Banach D-algebra** if any fundamental sequence of elements of algebra A_1 converges, i.e. has limit in algebra A_1.* □

DEFINITION 2.5.2. *A map*

$$f : A_1 \to A_2$$

*of Banach D_1-algebra A_1 with norm $|x|_1$ into Banach D_2-algebra A_2 with norm $|y|_2$ is called **continuous**, if for every as small as we please $\epsilon > 0$ there exist such $\delta > 0$, that*

$$|x' - x|_1 < \delta$$

implies

$$|f(x') - f(x)|_2 < \epsilon$$

□

DEFINITION 2.5.3. *Let A be Banach D-module with norm $\|x\|_A$. Let B be Banach D-module with norm $\|x\|_B$. For map*

$$f : A^n \to B$$

the value

(2.5.1) $$\|f\| = \sup \frac{\|f(a_1, ..., a_n)\|_B}{\|a_1\|_A ... \|a_n\|_A}$$

*is called **norm of map** f.* □

THEOREM 2.5.4. *For map*

$$f : A^n \to B$$

of Banach D-module A with norm $\|x\|_A$ to Banach D-module B with norm $\|x\|_B$

(2.5.2) $$\|f(a_1, ..., a_n)\|_B \leq \|f\| \|a_1\|_A ... \|a_n\|_A$$

PROOF. The theorem follows from the theorem [8]-3.2.19. □

THEOREM 2.5.5. *Let*
$$o_n : A^n \to B$$
be sequence of maps of Banach D-module A into Banach D-module B such that

(2.5.3) $$\lim_{n \to \infty} \|o_n\| = 0$$

Then, for any B-numbers $a_1, ..., a_p$,

(2.5.4) $$\lim_{n \to \infty} o_n(a_1, ..., a_p) = 0$$

PROOF. The theorem follows from the theorem [8]-3.2.20. □

DEFINITION 2.5.6. *Let A be Banach D-module with norm $\|a\|_A$. Let B be Banach D-module with norm $\|a\|_B$. The map*
$$f : A \to B$$
is called **differentiable** *on the set $U \subset A$, if, at every point $x \in U$, the increment of the map f can be represented as*

(2.5.5) $$f(x+h) - f(x) = d_x f(x) \circ h + o(h) = \frac{df(x)}{dx} \circ h + o(h)$$

where
$$\frac{df(x)}{dx} : A \to B$$
is linear map of D-module A into D-module B and
$$o : A \to B$$
is such continuous map that
$$\lim_{a \to 0} \frac{\|o(a)\|_B}{\|a\|_A} = 0$$

Linear map $\frac{df(x)}{dx}$ is called **derivative of map** f. □

DEFINITION 2.5.7. *Since, for given x, we consider the increment (2.5.5) of the map*
$$f : A \to A$$
as function of differential dx of variable x, then the linear part of this function
$$df = \frac{df(x)}{dx} \circ dx$$
is called **differential of map** f. □

REMARK 2.5.8. *According to definition 2.5.6, the derivative of the map f is the map*
$$x \in A \to \frac{df(x)}{dx} \in \mathcal{L}(D; A \to B)$$
Expressions $d_x f(x)$ and $\frac{df(x)}{dx}$ are different notations for the same map. □

THEOREM 2.5.9. *Definitions of the derivative* (2.5.5) *is equivalent to the definition*

(2.5.6) $$\frac{df(x)}{dx} \circ a = \lim_{t \to 0,\ t \in R} (t^{-1}(f(x+ta) - f(x)))$$

PROOF. The theorem follows from the theorem [8]-3.3.4. □

DEFINITION 2.5.10. *Polylinear map*

(2.5.7) $$\frac{d^2 f(x)}{dx^2} \circ (a_1; a_2) = d_{x^2}^2 f(x) \circ (a_1; a_2) = \frac{d}{dx}\left(\frac{df(x)}{dx} \circ a_1\right) \circ a_2$$

is called **derivative of second order** *of map* f. □

DEFINITION 2.5.11. *By induction, assuming that we defined the derivative* $\frac{d^{n-1} f(x)}{dx^{n-1}}$ *of order* $n-1$, *we define*

(2.5.8) $$\frac{d^n f(x)}{dx^n} \circ (a_1; ...; a_n) = d_{x^n}^n f(x) \circ (a_1; ...; a_n)$$
$$= \frac{d}{dx}\left(\frac{d^{n-1} f(x)}{dx^{n-1}} \circ (a_1; ...; a_{n-1})\right) \circ a_n$$

derivative of order n *of map* f.

We also assume $\frac{d^0 f(x)}{dx^0} = f(x)$. □

2.6. Quaternion Algebra

DEFINITION 2.6.1. *Let R be real field. Extension field $H = R(i,j,k)$ is called* **the quaternion algebra** *if multiplication in algebra H is defined according to rule*

(2.6.1)

	i	j	k
i	-1	k	$-j$
j	$-k$	-1	i
k	j	$-i$	-1

□

Elements of the algebra H have form
$$x = x^0 + x^1 i + x^2 j + x^3 k$$
$$x^0, x^1, x^2, x^3 \in R$$

Quaternion

(2.6.2) $$\overline{x} = x^0 - x^1 i - x^2 j - x^3 k$$

is called conjugate to the quaternion x. We define **the norm of the quaternion** x using equation

$$|x|^2 = x\overline{x} = (x^0)^2 + (x^1)^2 + (x^2)^2 + (x^3)^2 \qquad (2.6.3)$$

From the equality (2.6.3), it follows that inverse element has form

$$x^{-1} = |x|^{-2}\overline{x} \qquad (2.6.4)$$

THEOREM 2.6.2. *Let*

$$e_0 = 1 \quad e_1 = i \quad e_2 = j \quad e_3 = k \qquad (2.6.5)$$

be basis of quaternion algebra H. Then in the basis (2.6.5), *structural constants have form*

$$\begin{aligned}
C_{00}^0 &= 1 & C_{01}^1 &= 1 & C_{02}^2 &= 1 & C_{03}^3 &= 1 \\
C_{10}^1 &= 1 & C_{11}^0 &= -1 & C_{12}^3 &= 1 & C_{13}^2 &= -1 \\
C_{20}^2 &= 1 & C_{21}^3 &= -1 & C_{22}^0 &= -1 & C_{23}^1 &= 1 \\
C_{30}^3 &= 1 & C_{31}^2 &= 1 & C_{32}^1 &= -1 & C_{33}^0 &= -1
\end{aligned} \qquad (2.6.6)$$

PROOF. Value of structural constants follows from the multiplication table (2.6.1). □

2.7. Direct Sum of D-modules

DEFINITION 2.7.1. *Let $\mathcal{A}$ be a category. Let $\{B_i, i \in I\}$ be the set of objects of $\mathcal{A}$. Object*

$$P = \coprod_{i \in I} B_i$$

and set of morphisms

$$\{f_i : B_i \to P, i \in I\}$$

is called a **coproduct of set of objects** *$\{B_i, i \in I\}$ in category $\mathcal{A}$*[2.12] *if for any object R and set of morphisms*

$$\{g_i : B_i \to R, i \in I\}$$

there exists a unique morphism

$$h : P \to R$$

such that diagram

$$P \xleftarrow{f_i} B_i \qquad h \circ f_i = g_i$$
$$h \downarrow \swarrow g_i$$
$$R$$

is commutative for all $i \in I$.

[2.12] I made definition according to [1], page 59.

If $|I|=n$, then we also will use notation
$$P = \coprod_{i=1}^{n} B_i = B_1 \coprod ... \coprod B_n$$
for coproduct of set of objects $\{B_i, i \in I\}$ in $\mathcal{A}$. □

DEFINITION 2.7.2. *Coproduct in category of Abelian groups Ab is called* **direct sum**.[2.13] *We will use notation* $A \oplus B$ *for direct sum of Abelian groups A and B.* □

THEOREM 2.7.3. *Let $\{A_i, i \in I\}$ be set of Abelian groups. Let*
$$A \subseteq \prod_{i \in I} A_i$$
be such set that $(x_i, i \in I) \in A$, if $x_i \ne 0$ for finite number of indices i. Then[2.14]

(2.7.1) $$A = \bigoplus_{i \in I} A_i$$

PROOF. According to construction, A is subgroup of Abelian group $\prod A_i$. The map
$$\lambda_j : A_j \to A$$
defined by the equality

(2.7.2) $$\lambda_j(x) = (\delta_j^i x, i \in I)$$

is an injective homomorphism.

Let
$$\{f_i : A_i \to B, i \in I\}$$
be set of homomorphisms into Abelian group B. We define the map
$$f : A \to B$$
by the equality

(2.7.3) $$f\left(\bigoplus_{i \in I} x_i\right) = \sum_{i \in I} f_i(x_i)$$

The sum in the right side of the equality (2.7.3) is finite, since all summands, except for a finite number, equal 0. From the equality
$$f_i(x_i + y_i) = f_i(x_i) + f_i(y_i)$$

[2.13] See also definition in [1], pages 36, 37.
[2.14] See also proposition [1]-7.1, page 37.

and the equality (2.7.3), it follows that

$$f(\bigoplus_{i\in I}(x_i + y_i)) = \sum_{i\in I} f_i(x_i + y_i) = \sum_{i\in I}(f_i(x_i) + f_i(y_i))$$
$$= \sum_{i\in I} f_i(x_i) + \sum_{i\in I} f_i(y_i)$$
$$= f(\bigoplus_{i\in I} x_i) + f(\bigoplus_{i\in I} y_i)$$

Therefore, the map f is homomorphism of Abelian group. The equality

$$f \circ \lambda_j(x) = \sum_{i\in I} f_i(\delta^i_j x) = f_j(x)$$

follows from equalities (2.7.2), (2.7.3). Since the map λ_i is injective, then the map f is unique. Therefore, the theorem folows from definitions 2.7.1, 2.7.2. □

THEOREM 2.7.4. *Direct sum of Abelian groups A_1, ..., A_n coincides with their Cartesian product*

$$A_1 \oplus ... \oplus A_n = A_1 \times ... \times A_n$$

PROOF. The theorem follows from the theorem 2.7.3. □

Let

$$A = A_1 \oplus ... \oplus A_n$$

be direct sum of Abelian groups A_1, ..., A_n. According to the proof of the theorem 2.7.3, any A-number a has form $(a_1, ..., a_n)$ where $a_i \in A_i$. We also will use notation

$$a = a_1 \oplus ... \oplus a_n$$

DEFINITION 2.7.5. *Coproduct in category of D-modules is called* **direct sum**.[2.15] *We will use notation $A \oplus B$ for direct sum of D-modules A and B.* □

THEOREM 2.7.6. *Let $\{A_i, i \in I\}$ be set of D-modules. Then the representation*

$$D \dashrightarrow \bigoplus_{i\in I} A_i \qquad d\left(\bigoplus_{i\in I} a_i\right) = \bigoplus_{i\in I} d a_i$$

of the ring D in direct sum of Abelian groups

$$A = \bigoplus_{i\in I} A_i$$

is direct sum of D-modules

$$A = \bigoplus_{i\in I} A_i$$

[2.15] See also definition in [1], pages 36, 37.

PROOF.
Let
$$\{f_i : A_i \to B, i \in I\}$$
be set of linear maps into D-module B. We define the map
$$f : A \to B$$
by the equality

(2.7.4)
$$f\left(\bigoplus_{i \in I} x_i\right) = \sum_{i \in I} f_i(x_i)$$

The sum in the right side of the equality (2.7.4) is finite, since all summands, except for a finite number, equal 0. From the equality
$$f_i(x_i + y_i) = f_i(x_i) + f_i(y_i)$$
and the equality (2.7.4), it follows that

$$f(\bigoplus_{i \in I}(x_i + y_i)) = \sum_{i \in I} f_i(x_i + y_i) = \sum_{i \in I}(f_i(x_i) + f_i(y_i))$$
$$= \sum_{i \in I} f_i(x_i) + \sum_{i \in I} f_i(y_i)$$
$$= f(\bigoplus_{i \in I} x_i) + f(\bigoplus_{i \in I} y_i)$$

From the equality
$$f_i(dx_i) = df_i(x_i)$$
and the equality (2.7.4), it follows that
$$f((dx_i, i \in I)) = \sum_{i \in I} f_i(dx_i) = \sum_{i \in I} df_i(x_i) = d \sum_{i \in I} f_i(x_i)$$
$$= df((x_i, i \in I))$$

Therefore, the map f is linear map. The equality
$$f \circ \lambda_j(x) = \sum_{i \in I} f_i(\delta^i_j x) = f_j(x)$$
follows from equalities (2.7.2), (2.7.4). Since the map λ_i is injective, then the map f is unique. Therefore, the theorem folows from definitions 2.7.1, 2.7.2. □

THEOREM 2.7.7. *Direct sum of D-modules A_1, ..., A_n coincides with their Cartesian product*
$$A_1 \oplus ... \oplus A_n = A_1 \times ... \times A_n$$

PROOF. The theorem follows from the theorem 2.7.6. □

THEOREM 2.7.8. *Let* A^1, ..., A^n *be D-modules and*
$$A = A^1 \oplus ... \oplus A^n$$
Let us represent A-number
$$a = a^1 \oplus ... \oplus a^n$$
as column vector
$$a = \begin{pmatrix} a^1 \\ ... \\ a^n \end{pmatrix}$$
Let us represent a linear map
$$f : A \to B$$
as row vector
$$f = \begin{pmatrix} f_1 & ... & f_n \end{pmatrix}$$
$$f_i : A^i \to B$$
Then we can represent value of the map f in A-number a as product of matrices

(2.7.5) $$f \circ a = \begin{pmatrix} f_1 & ... & f_n \end{pmatrix} \circ^\circ \begin{pmatrix} a^1 \\ ... \\ a^n \end{pmatrix} = f_i \circ a^i$$

PROOF. The theorem follows from the definition (2.7.4). □

THEOREM 2.7.9. *Let* B^1, ..., B^m *be D-modules and*
$$B = B^1 \oplus ... \oplus B^m$$
Let us represent B-number
$$b = b^1 \oplus ... \oplus b^m$$
as column vector
$$b = \begin{pmatrix} b^1 \\ ... \\ b^m \end{pmatrix}$$
Then the linear map
$$f : A \to B$$
has representation as column vector of maps
$$f = \begin{pmatrix} f^1 \\ ... \\ f^m \end{pmatrix}$$

such way that, if $b = f \circ a$, then

$$\begin{pmatrix} b^1 \\ ... \\ b^m \end{pmatrix} = \begin{pmatrix} f^1 \\ ... \\ f^m \end{pmatrix} \circ a = \begin{pmatrix} f^1 \circ a \\ ... \\ f^m \circ a \end{pmatrix}$$

PROOF. The theorem follows from the theorem [10]-2.1.5. □

THEOREM 2.7.10. *Let* A^1, ..., A^n, B^1, ..., B^m *be D-modules and*
$$A = A^1 \oplus ... \oplus A^n$$
$$B = B^1 \oplus ... \oplus B^m$$

Let us represent A-number
$$a = a^1 \oplus ... \oplus a^n$$
as column vector
$$a = \begin{pmatrix} a^1 \\ ... \\ a^n \end{pmatrix}$$

Let us represent B-number
$$b = b^1 \oplus ... \oplus b^m$$
as column vector
$$b = \begin{pmatrix} b^1 \\ ... \\ b^m \end{pmatrix}$$

Then the linear map f has representation as a matrix of maps
$$f = \begin{pmatrix} f_1^1 & ... & f_n^1 \\ ... & ... & ... \\ f_1^m & ... & f_n^m \end{pmatrix}$$
such way that, if $b = f \circ a$, then

(2.7.6)
$$\begin{pmatrix} b^1 \\ ... \\ b^m \end{pmatrix} = \begin{pmatrix} f_1^1 & ... & f_n^1 \\ ... & ... & ... \\ f_1^m & ... & f_n^m \end{pmatrix} \circ \begin{pmatrix} a^1 \\ ... \\ a^n \end{pmatrix} = \begin{pmatrix} f_i^1 \circ a^i \\ ... \\ f_i^m \circ a^i \end{pmatrix}$$

The map
$$f_j^i : A^j \to B^i$$
is a linear map and is called **partial linear map**.

PROOF. According to the theorem 2.7.9, there exists the set of linear maps
$$f^i : A \to B^i$$
such that

(2.7.7) $$\begin{pmatrix} b^1 \\ ... \\ b^m \end{pmatrix} = \begin{pmatrix} f^1 \\ ... \\ f^m \end{pmatrix} \circ a = \begin{pmatrix} f^1 \circ a \\ ... \\ f^m \circ a \end{pmatrix}$$

According to the theorem 2.7.8, for every i, there exists the set of linear maps
$$f^i_j : A^j \to B^i$$
such that

(2.7.8) $$f^i \circ a = \begin{pmatrix} f^i_1 & ... & f^i_n \end{pmatrix} \circ \begin{pmatrix} a^1 \\ ... \\ a^n \end{pmatrix} = f^i_j \circ a^j$$

If we identify matrices
$$\begin{pmatrix} \begin{pmatrix} f^1_1 & ... & f^1_n \end{pmatrix} \\ ... \\ \begin{pmatrix} f^m_1 & ... & f^m_n \end{pmatrix} \end{pmatrix} = \begin{pmatrix} f^1_1 & ... & f^1_n \\ ... & ... & ... \\ f^m_1 & ... & f^m_n \end{pmatrix}$$

then the equality (2.7.6) follows from equalities (2.7.7), (2.7.8). □

Let B^1, ..., B^m be D-algebras. Then we can represent linear map f^i_j using $B^i \otimes B^i$-number.

2.8. Direct Sum of Banach D-Modules

THEOREM 2.8.1. *Let A^1, ..., A^n be Banach D-modules and*
$$A = A^1 \oplus ... \oplus A^n$$
Then, in D-module A, we can introduce norm such that D-module A is Banach D-module.

PROOF. Let $\|a^i\|_i$ be norm in D-module A^i.
2.8.1.1: We introduce norm in D-module A by the equality
$$\|b\| = \max(\|b^i\|_i, i = 1, ..., n)$$
where
$$b = b^1 \oplus ... \oplus b^n$$
2.8.1.2: Let $\{a_p\}$, $p = 1, ...,$ be fundamental sequence where
$$a_p = a^1_p \oplus ... \oplus a^n_p$$

2.8. Direct Sum of Banach D-Modules

2.8.1.3: Therefore, for any $\epsilon \in R$, $\epsilon > 0$, there exists N such that for any p, $q > N$
$$\|a_p - a_q\| < \epsilon$$

2.8.1.4: According to statements 2.8.1.1, 2.8.1.2, 2.8.1.3,
$$\|a_p^i - a_q^i\|_i < \epsilon$$
for any $p, q > N$ and $i = 1, ..., n$.

2.8.1.5: Therefore, the sequence $\{a_p^i\}$, $i = 1, ..., n$, $p = 1, ...,$ is fundamental sequence in D-module A^i and there exists limit
$$a^i = \lim_{p \to \infty} a_p^i$$

2.8.1.6: Let
$$a = a^1 \oplus ... \oplus a^n$$

2.8.1.7: According to the statement 2.8.1.5, for any $\epsilon \in R$, $\epsilon > 0$, there exists N_i such that for any $p > N_i$
$$\|a^i - a_p^i\|_i < \epsilon$$

2.8.1.8: Let
$$N = \max(N_1, ..., N_n)$$

2.8.1.9: According to statements 2.8.1.6, 2.8.1.7, 2.8.1.8, for any $\epsilon \in R$, $\epsilon > 0$, there exists N such that for any $p > N$
$$\|a - a_p\|_i < \epsilon$$

2.8.1.10: Therefore,
$$a = \lim_{p \to \infty} a_p$$

The theorem follows from statements 2.8.1.1, 2.8.1.2, 2.8.1.10. □

Using the theorem 2.8.1, we can consider the derivative of a map
$$f : A^1 \oplus ... \oplus A^n \to B^1 \oplus ... \oplus B^m$$

THEOREM 2.8.2. *Let $A^1, ..., A^n$, $B^1, ..., B^m$ be Banach D-modules and*
$$A = A^1 \oplus ... \oplus A^n$$
$$B = B^1 \oplus ... \oplus B^m$$

Let us represent differential
$$dx = dx^1 \oplus ... \oplus dx^n$$

as column vector
$$dx = \begin{pmatrix} dx^1 \\ ... \\ dx^n \end{pmatrix}$$

Let us represent differential
$$dy = dy^1 \oplus ... \oplus dy^m$$

as column vector

$$dy = \begin{pmatrix} dy^1 \\ ... \\ dy^m \end{pmatrix}$$

Then the derivative of the map

$$f : A \to B$$
$$f = f^1 \oplus ... \oplus f^m$$

has representation

$$\frac{df}{dx} = \begin{pmatrix} \dfrac{\partial f^1}{\partial x^1} & ... & \dfrac{\partial f^1}{\partial x^n} \\ ... & ... & ... \\ \dfrac{\partial f^m}{\partial x^1} & ... & \dfrac{\partial f^m}{\partial x^n} \end{pmatrix}$$

such way that

(2.8.1) $$\begin{pmatrix} dy^1 \\ ... \\ dy^m \end{pmatrix} = \begin{pmatrix} \dfrac{\partial f^1}{\partial x^1} & ... & \dfrac{\partial f^1}{\partial x^n} \\ ... & ... & ... \\ \dfrac{\partial f^m}{\partial x^1} & ... & \dfrac{\partial f^m}{\partial x^n} \end{pmatrix} \circ \begin{pmatrix} dx^1 \\ ... \\ dx^n \end{pmatrix} = \begin{pmatrix} \dfrac{\partial f^1}{\partial x^i} \circ dx^i \\ ... \\ \dfrac{\partial f^m}{\partial x^i} \circ dx^i \end{pmatrix}$$

STATEMENT 2.8.3. *The linear map* $\dfrac{\partial f^i}{\partial x^j}$ *is called* **partial derivative** *and this map is the derivative of map f^i with respect to variable x^j assuming that other coordinates of A-number x are fixed.* ⊙

PROOF. The equality (2.8.1) follows from the equality (2.7.6).
We can represent the map

$$f^i : A \to B^i$$

as

$$f^i(x) = f^i(x^1, ..., x^n)$$

The equality

(2.8.2) $$\frac{df^i(x)}{dx} \circ dx = \frac{\partial f^i(x^1, ..., x^n)}{\partial x^j} \circ dx^j$$

follows from the equality (2.8.1). According to the theorem 2.5.9,

$$
\begin{aligned}
(2.8.3)\quad \frac{df^i(x)}{dx} \circ dx &= \lim_{t\to 0,\ t\in R}(t^{-1}(f^(x+tdx) - f(x)))\\
&= \lim_{t\to 0,\ t\in R}(t^{-1}(f^i(x^1+tdx^1, x^2+tdx^2, ..., x^n+tdx^n)\\
&\quad - f^i(x^1, x^2+tdx^2, ..., x^n+tdx^n)\\
&\quad + f^i(x^1, x^2+tdx^2, ..., x^n+tdx^n) - ...\\
&\quad - f^i(x^1, x^2, ..., x^n)))\\
&= \lim_{t\to 0,\ t\in R}(t^{-1}(f^i(x^1+tdx^1, x^2+tdx^2, ..., x^n+tdx^n)\\
&\quad - f^i(x^1, x^2+tdx^2, ..., x^n+tdx^n))) + ...\\
&\quad + \lim_{t\to 0,\ t\in R}(t^{-1}(f^i(x^1, ..., x^n+tdx^n) - f^i(x^1, ..., x^n)))\\
&= f^i_1 \circ dx^1 + ... + f^i_n \circ dx^n
\end{aligned}
$$

where f^i_j is the derivative of map f^i with respect to variable x^j assuming that other coordinates of A-number x are fixed. The equality

$$(2.8.4)\qquad f^i_j = \frac{\partial f^i(x^1, ..., x^n)}{\partial x^j}$$

follows from equalities (2.8.2), (2.8.3). The statement 2.8.3 follows from the equality (2.8.4). □

EXAMPLE 2.8.4. *Consider map*

$$(2.8.5)\quad \begin{aligned} y^1 &= f^1(x^1, x^2, x^3) = (x^1)^2 + x^2 x^3\\ y^2 &= f^2(x^1, x^2, x^3) = x^1 x^2 + (x^3)^2 \end{aligned}$$

Therefore

$$\frac{\partial y^1}{\partial x^1} = x^1 \otimes 1 + 1 \otimes x^1 \qquad \frac{\partial y^1}{\partial x^2} = 1 \otimes x^3 \qquad \frac{\partial y^1}{\partial x^3} = x^2 \otimes 1$$

$$\frac{\partial y^2}{\partial x^1} = 1 \otimes x^2 \qquad \frac{\partial y^2}{\partial x^2} = x^1 \otimes 1 \qquad \frac{\partial y^2}{\partial x^3} = x^3 \otimes 1 + 1 \otimes x^3$$

and the derivative of the map (2.8.5) *is*

$$(2.8.6)\qquad \frac{df}{dx} = \begin{pmatrix} x^1 \otimes 1 + 1 \otimes x^1 & 1 \otimes x^3 & x^2 \otimes 1\\ 1 \otimes x^2 & x^1 \otimes 1 & x^3 \otimes 1 + 1 \otimes x^3 \end{pmatrix}$$

The equality

$$(2.8.7)\quad \begin{aligned} dy^1 &= (x^1 \otimes 1 + 1 \otimes x^1) \circ dx^1 + (1 \otimes x^3) \circ dx^2 + (x^2 \otimes 1) \circ dx^3\\ &= x^1 dx^1 + dx^1 x^1 + dx^2 x^3 + x^2 dx^3\\ dy^2 &= (1 \otimes x^2) \circ dx^1 + (x^1 \otimes 1) \circ dx^2 + (x^3 \otimes 1 + 1 \otimes x^3) \circ dx^3\\ &= dx^1 x^2 + x^1 dx^2 + x^3 dx^3 + dx^3 x^3 \end{aligned}$$

follows from the equality (2.8.6). We also can get the expression (2.8.7) by direct calculation

$$\begin{aligned}
dy^1 &= f^1(x+dx) - f^1(x) \\
&= (x^1+dx^1)^2 + (x^2+dx^2)(x^3+dx^3) - (x^1)^2 - x^2 x^3 \\
&= (x^1)^2 + x^1 dx^1 + dx^1 x^1 + x^2 x^3 + dx^2 x^3 + x^2 dx^3 - (x^1)^2 - x^2 x^3 \\
&= x^1 dx^1 + dx^1 x^1 + dx^2 x^3 + x^2 dx^3
\end{aligned}$$

(2.8.8)
$$\begin{aligned}
dy^2 &= f^2(x+dx) - f^2(x) \\
&= (x^1+dx^1)(x^2+dx^2) + (x^3+dx^3)^2 - x^1 x^2 - (x^3)^2 \\
&= x^1 x^2 + dx^1 x^2 + x^1 dx^2 + (x^3)^2 + x^3 dx^3 + dx^3 x^3 - x^1 x^2 - (x^3)^2 \\
&= dx^1 x^2 + x^1 dx^2 + x^3 dx^3 + dx^3 x^3
\end{aligned}$$

□

THEOREM 2.8.5. *Let A^1, ..., A^n, B be Banach D-modules and*

$$A = A^1 \oplus ... \oplus A^n$$

If the map

$$f : A \to B$$

has the second derivative, then the second derivative has the following form

$$\frac{d^2 f}{dx^2} \circ (h_1, h_2) = \frac{\partial^2 f}{\partial x^i \partial x^j} \circ (h_1^i, h_2^j)$$

where

$$h_1 = h_1^1 \oplus ... \oplus h_1^n$$
$$h_2 = h_2^1 \oplus ... \oplus h_2^n$$

and we define **partial derivative of second order** *by the equality*

$$\frac{\partial^2 f}{\partial x^i \partial x^j} = \frac{\partial}{\partial x^i} \frac{\partial f}{\partial x^j}$$

PROOF. The theorem follows from the definition 2.5.10 and from the theorem 2.8.2. □

THEOREM 2.8.6. *Let A^1, ..., A^n, B be Banach D-modules and*

$$A = A^1 \oplus ... \oplus A^n$$

Let derivatives of map

$$f : A \to B$$

are continuous and differentiable on the set $U \subset A$. Let partial derivatives of second order are continuous on the set $U \subset A$. Then on the set U partial derivatives satisfy equality

(2.8.9)
$$\frac{\partial^2 f(x)}{\partial x^j \partial x^i} \circ (h_1^j, h_2^i) = \frac{\partial^2 f(x)}{\partial x^i \partial x^j} \circ (h_1^i, h_2^j)$$

PROOF. The theorem follows from the theorem [5]-9.1.6. □

CHAPTER 3

Differential Equation Solved with Respect to the Derivative

3.1. Differential Equation $\dfrac{dy}{dx} = 1 \otimes x + x \otimes 1$

DEFINITION 3.1.1. *Let A, B be Banach D-modules. The map*
$$g : A \to \mathcal{L}(D; A \to B)$$
is called **integrable***, if there exists a map*
$$f : A \to B$$
such that
$$\frac{df(x)}{dx} = g(x)$$
Then we use notation
$$f(x) = \int g(x) \circ dx$$
and the map f is called **indefinite integral** *of the map g.* □

It is convenient to use table of derivatives [3.1] and table of integrals [3.2] to solve differential equation
$$\frac{dy}{dx} = g(x)$$
if the map g is simple.

DEFINITION 3.1.2. *If there exist indefinite integral*
$$\int g(x) \circ dx$$
then differential equation
$$\frac{dy}{dx} = g(x)$$
is called **integrable***.* □

[3.1] In case of maps of real field, see for instance sections [16]-95, [2]-2.3. In case of maps of Banach algebra, see also section A.1.

[3.2] In case of maps of real field, see for instance sections [17]-265, [2]-4.4. In case of maps of Banach algebra, see also section A.2.

EXAMPLE 3.1.3. *Let A be Banach algebra. According to the theorem A.2.7, the map*
$$y = x^2 + C$$
is solution of the differential equation

(3.1.1) $$\frac{dy}{dx} = x \otimes 1 + 1 \otimes x$$

with initial condition
$$x_0 = 0 \quad y_0 = C$$

□

If D-algebra B has finite basis, then we can write the differential equation
$$\frac{dy}{dx} = g(x)$$
as system of differential equations
$$\frac{\partial y^i}{\partial x^j} = g^i_j \quad y = y^i e_{B \cdot i} \quad x = x^i e_{A \cdot i}$$
In order that the calculations do not hide idea how to make transformation, we consider relatively simple example 3.1.3.

We start with the differential equation (3.1.1) in complex field. The theorems 3.1.4, 3.1.6 are simple. But they are necessary to prepare the reader for the proof of the theorem 3.1.10.

THEOREM 3.1.4. *The differential equation (3.1.1) in complex field has form of system of differential equations*

(3.1.2) $$\begin{cases} \dfrac{\partial y^0}{\partial x^0} = 2x^0 & \dfrac{\partial y^0}{\partial x^1} = -2x^1 \\ \dfrac{\partial y^1}{\partial x^0} = 2x^1 & \dfrac{\partial y^1}{\partial x^1} = 2x^0 \end{cases}$$

with respect to the basis
$$e_{C \cdot 0} = 1 \quad e_{C \cdot 1} = i$$

PROOF. Since product in complex field is commutative, we can write differential equation (3.1.1) as follows

(3.1.3) $$dy = 2x\,dx$$

If we represent differentials dx, dy as vector-column, then, according to the theorem [6]-2.5.1, the equation (3.1.3) gets following form

(3.1.4) $$\begin{pmatrix} dy^0 \\ dy^1 \end{pmatrix} = \begin{pmatrix} 2x^0 & -2x^1 \\ 2x^1 & 2x^0 \end{pmatrix} \begin{pmatrix} dx^0 \\ dx^1 \end{pmatrix}$$

3.1. Differential Equation $\frac{dy}{dx} = 1 \otimes x + x \otimes 1$

Since the matrix of derivative has following form

$$\frac{dy}{dx} = \begin{pmatrix} \frac{\partial y^0}{\partial x^0} & \frac{\partial y^0}{\partial x^1} \\ \frac{\partial y^1}{\partial x^0} & \frac{\partial y^1}{\partial x^1} \end{pmatrix}$$

then the equality

(3.1.5) $$\begin{pmatrix} \frac{\partial y^0}{\partial x^0} & \frac{\partial y^0}{\partial x^1} \\ \frac{\partial y^1}{\partial x^0} & \frac{\partial y^1}{\partial x^1} \end{pmatrix} = \begin{pmatrix} 2x^0 & -2x^1 \\ 2x^1 & 2x^0 \end{pmatrix}$$

follows from (3.1.4). The system of differential equations (3.1.15) follows from the equality (3.1.5). □

LEMMA 3.1.5. *Let*
$$x = x^0 + x^1 i$$
be complex number. Then

(3.1.6) $$x^2 = (x^0)^2 - (x^1)^2 + 2x^0 x^1 i$$

PROOF. The equality (3.1.6) follows from the definition of product in complex field. □

THEOREM 3.1.6. *The map*
$$y = x^2 + C$$
is solution of the system of differential equations (3.1.2) *in complex field with initial condition*
$$x = 0 \quad y^0 = C^0 \quad y^1 = C^1 \quad C = C^0 + C^1 i$$

PROOF. The map

(3.1.7) $$y^0 = (x^0)^2 + C_1^0(x^1)$$

is solution of the differential equation
$$\frac{\partial y^0}{\partial x^0} = 2x^0$$

From (3.1.2), (3.1.7), it follows that

(3.1.8) $$\frac{\partial y^0}{\partial x^1} = \frac{dC_1^0}{dx^1} = -2x^1$$

The map

(3.1.9) $$C_1^0(x^1) = -(x^1)^2 + C^0$$

is solution of the differential equation (3.1.8). The equality

(3.1.10) $$y^0 = (x^0)^2 - (x^1)^2 + C^0$$

follows from equalities (3.1.7), (3.1.9).

The map
$$(3.1.11) \qquad y^1 = 2x^0 x^1 + C_1^1(x^1)$$
is solution of the differential equation
$$\frac{\partial y^1}{\partial x^0} = 2x^1$$
From (3.1.2), (3.1.11), it follows that
$$(3.1.12) \qquad \frac{\partial y^1}{\partial x^1} = 2x^0 + \frac{dC_1^1}{dx^1} = 2x^0$$
The map
$$(3.1.13) \qquad C_1^1(x^1) = C^1$$
is solution of the differential equation (3.1.12). The equality
$$(3.1.14) \qquad y^1 = 2x^0 x^1 + C^1$$
follows from equalities (3.1.11), (3.1.13).

The theorem follows from the comparison of equality (3.1.6) and equalities (3.1.10), (3.1.14). □

We consider the differential equation (3.1.1) in quaternion algebra in the same way as we considered this differential equation in complex field.

THEOREM 3.1.7. *The differential equation (3.1.1) in quaternion algebra has form of system of differential equations*

$$(3.1.15) \quad \begin{cases} \dfrac{\partial y^0}{\partial x^0} = 2x^0 & \dfrac{\partial y^0}{\partial x^1} = -2x^1 & \dfrac{\partial y^0}{\partial x^2} = -2x^2 & \dfrac{\partial y^0}{\partial x^3} = -2x^3 \\[4pt] \dfrac{\partial y^1}{\partial x^0} = 2x^1 & \dfrac{\partial y^1}{\partial x^1} = 2x^0 & \dfrac{\partial y^1}{\partial x^2} = 0 & \dfrac{\partial y^1}{\partial x^3} = 0 \\[4pt] \dfrac{\partial y^2}{\partial x^0} = 2x^2 & \dfrac{\partial y^2}{\partial x^1} = 0 & \dfrac{\partial y^2}{\partial x^2} = 2x^0 & \dfrac{\partial y^2}{\partial x^3} = 0 \\[4pt] \dfrac{\partial y^3}{\partial x^0} = 2x^3 & \dfrac{\partial y^3}{\partial x^1} = 0 & \dfrac{\partial y^3}{\partial x^2} = 0 & \dfrac{\partial y^3}{\partial x^3} = 2x^0 \end{cases}$$

with respect to the basis
$$e_0 = 1 \quad e_1 = i \quad e_2 = j \quad e_3 = k$$

PROOF. We can write differential equation (3.1.1) as follows
$$(3.1.16) \qquad dy = x\,dx + dx\,x$$

3.1. Differential Equation $\frac{dy}{dx} = 1 \otimes x + x \otimes 1$

If we represent differentials dx, dy as vector-column, then, according to theorems [7]-5.1, [7]-5.2, the equation (3.1.16) gets following form

$$\begin{pmatrix} dy^0 \\ dy^1 \\ dy^2 \\ dy^3 \end{pmatrix} = E_l(x) \begin{pmatrix} dx^0 \\ dx^1 \\ dx^2 \\ dx^3 \end{pmatrix} + E_r(x) \begin{pmatrix} dx^0 \\ dx^1 \\ dx^2 \\ dx^3 \end{pmatrix}$$

(3.1.17)

$$= \begin{pmatrix} x^0 & -x^1 & -x^2 & -x^3 \\ x^1 & x^0 & -x^3 & x^2 \\ x^2 & x^3 & x^0 & -x^1 \\ x^3 & -x^2 & x^1 & x^0 \end{pmatrix} \begin{pmatrix} dx^0 \\ dx^1 \\ dx^2 \\ dx^3 \end{pmatrix}$$

$$+ \begin{pmatrix} x^0 & -x^1 & -x^2 & -x^3 \\ x^1 & x^0 & x^3 & -x^2 \\ x^2 & -x^3 & x^0 & x^1 \\ x^3 & x^2 & -x^1 & x^0 \end{pmatrix} \begin{pmatrix} dx^0 \\ dx^1 \\ dx^2 \\ dx^3 \end{pmatrix}$$

$$= \begin{pmatrix} 2x^0 & -2x^1 & -2x^2 & -2x^3 \\ 2x^1 & 2x^0 & 0 & 0 \\ 2x^2 & 0 & 2x^0 & 0 \\ 2x^3 & 0 & 0 & 2x^0 \end{pmatrix} \begin{pmatrix} dx^0 \\ dx^1 \\ dx^2 \\ dx^3 \end{pmatrix}$$

Since the matrix of derivative has following form

$$\frac{dy}{dx} = \begin{pmatrix} \frac{\partial y^0}{\partial x^0} & \frac{\partial y^0}{\partial x^1} & \frac{\partial y^0}{\partial x^2} & \frac{\partial y^0}{\partial x^3} \\ \frac{\partial y^1}{\partial x^0} & \frac{\partial y^1}{\partial x^1} & \frac{\partial y^1}{\partial x^2} & \frac{\partial y^1}{\partial x^3} \\ \frac{\partial y^2}{\partial x^0} & \frac{\partial y^2}{\partial x^1} & \frac{\partial y^2}{\partial x^2} & \frac{\partial y^2}{\partial x^3} \\ \frac{\partial y^3}{\partial x^0} & \frac{\partial y^3}{\partial x^1} & \frac{\partial y^3}{\partial x^2} & \frac{\partial y^3}{\partial x^3} \end{pmatrix}$$

then the equality

(3.1.18)
$$\begin{pmatrix} \frac{\partial y^0}{\partial x^0} & \frac{\partial y^0}{\partial x^1} & \frac{\partial y^0}{\partial x^2} & \frac{\partial y^0}{\partial x^3} \\ \frac{\partial y^1}{\partial x^0} & \frac{\partial y^1}{\partial x^1} & \frac{\partial y^1}{\partial x^2} & \frac{\partial y^1}{\partial x^3} \\ \frac{\partial y^2}{\partial x^0} & \frac{\partial y^2}{\partial x^1} & \frac{\partial y^2}{\partial x^2} & \frac{\partial y^2}{\partial x^3} \\ \frac{\partial y^3}{\partial x^0} & \frac{\partial y^3}{\partial x^1} & \frac{\partial y^3}{\partial x^2} & \frac{\partial y^3}{\partial x^3} \end{pmatrix} = \begin{pmatrix} 2x^0 & -2x^1 & -2x^2 & -2x^3 \\ 2x^1 & 2x^0 & 0 & 0 \\ 2x^2 & 0 & 2x^0 & 0 \\ 2x^3 & 0 & 0 & 2x^0 \end{pmatrix}$$

follows from (3.1.17). The system of differential equations (3.1.15) follows from the equality (3.1.18). □

LEMMA 3.1.8. *Let*

(3.1.19)
$$x = x^0 + x^1 i + x^2 j + x^3 k$$
$$y = y^0 + y^1 i + y^2 j + y^3 k$$

be quaternions. Then

(3.1.20)
$$xy = x^0 y^0 - x^1 y^1 - x^2 y^2 - x^3 y^3 + (x^0 y^1 + x^1 y^0 + x^2 y^3 - x^3 y^2)i$$
$$+ (x^0 y^2 + x^2 y^0 + x^3 y^1 - x^1 y^3)j + (x^0 y^3 + x^3 y^0 + x^1 y^2 - x^2 y^1)k$$

PROOF. The lemma follows from the definition 2.6.1. □

LEMMA 3.1.9. *Let*
$$x = x^0 + x^1 i + x^2 j + x^3 k$$
be quaternion. Then

(3.1.21) $\quad x^2 = (x^0)^2 - (x^1)^2 - (x^2)^2 - (x^3)^2 + 2x^0 x^1 i + 2x^0 x^2 j + 2x^0 x^3 k$

PROOF. The equality

(3.1.22)
$$x^2 = (x^0)^2 - (x^1)^2 - (x^2)^2 - (x^3)^2 + (x^0 x^1 + x^1 x^0 + x^2 x^3 - x^3 x^2)i$$
$$+ (x^0 x^2 + x^2 x^0 + x^3 x^1 - x^1 x^3)j + (x^0 x^3 + x^3 x^0 + x^1 x^2 - x^2 x^1)k$$

follows from the equality (3.1.20). The equality (3.1.21) follows from the equality (3.1.22). □

THEOREM 3.1.10. *The map*
$$y = x^2 + C$$
is solution of the system of differential equations (3.1.15) *with initial condition*
$$x = 0 \quad y^0 = C^0 \quad y^1 = C^1 \quad y^2 = C^2 \quad y^3 = C^3 \quad C = C^0 + C^1 i + C^2 j + C^3 k$$

PROOF. The map

(3.1.23) $\quad y^0 = (x^0)^2 + C_1^0(x^1, x^2, x^3)$

is solution of the differential equation
$$\frac{\partial y^0}{\partial x^0} = 2x^0$$
From (3.1.15), (3.1.23), it follows that

(3.1.24) $$\frac{\partial y^0}{\partial x^1} = \frac{\partial C_1^0}{\partial x^1} = -2x^1$$

The map

(3.1.25) $$C_1^0(x^1, x^2, x^3) = -(x^1)^2 + C_2^0(x^2, x^3)$$

is solution of the differential equation (3.1.24). The equality

(3.1.26) $$y^0 = (x^0)^2 - (x^1)^2 + C_2^0(x^2, x^3)$$

follows from equalities (3.1.23), (3.1.25). From (3.1.15), (3.1.26), it follows that

(3.1.27) $$\frac{\partial y^0}{\partial x^2} = \frac{\partial C_2^0}{\partial x^2} = -2x^2$$

The map

(3.1.28) $$C_2^0(x^2, x^3) = -(x^2)^2 + C_3^0(x^3)$$

is solution of the differential equation (3.1.27). The equality

(3.1.29) $$y^0 = (x^0)^2 - (x^1)^2 - (x^2)^2 + C_3^0(x^3)$$

follows from equalities (3.1.26), (3.1.28). From (3.1.15), (3.1.29), it follows that

(3.1.30) $$\frac{\partial y^0}{\partial x^3} = \frac{dC_3^0}{dx^3} = -2x^3$$

The map

(3.1.31) $$C_3^0(x^3) = -(x^3)^2 + C^0$$

is solution of the differential equation (3.1.30). The equality

(3.1.32) $$y^0 = (x^0)^2 - (x^1)^2 - (x^2)^2 - (x^3)^2 + C^0$$

follows from equalities (3.1.29), (3.1.31).

The map

(3.1.33) $$y^1 = 2x^0 x^1 + C_1^1(x^1, x^2, x^3)$$

is solution of the differential equation
$$\frac{\partial y^1}{\partial x^0} = 2x^1$$

From (3.1.15), (3.1.33), it follows that

(3.1.34) $$\frac{\partial y^1}{\partial x^1} = 2x^0 + \frac{\partial C_1^1}{\partial x^1} = 2x^0$$

The map

(3.1.35) $$C_1^1(x^1, x^2, x^3) = C_2^1(x^2, x^3)$$

is solution of the differential equation (3.1.34). The equality

(3.1.36) $$y^1 = 2x^0 x^1 + C_2^1(x^2, x^3)$$

follows from equalities (3.1.33), (3.1.35). From (3.1.15), (3.1.36), it follows that

(3.1.37) $$\frac{\partial y^1}{\partial x^2} = \frac{\partial C_2^1}{\partial x^2} = 0 \quad \frac{\partial y^1}{\partial x^3} = \frac{\partial C_2^1}{\partial x^3} = 0$$

The map

(3.1.38) $$C_2^1(x^2, x^3) = C^1$$

is solution of the system of differential equations (3.1.37). The equality

(3.1.39) $$y^1 = 2x^0 x^1 + C^1$$

follows from equalities (3.1.36), (3.1.38).

The similar way we see that

(3.1.40) $$y^2 = 2x^0 x^2 + C^2$$

(3.1.41) $$y^3 = 2x^0 x^3 + C^3$$

The theorem follows from the comparison of equality (3.1.21) and equalities (3.1.32), (3.1.39), (3.1.40), (3.1.41). □

3.2. Differential Equation $\frac{dy}{dx} = 3x \otimes x$

In this section I consider the differential equation in which existence of a solution depends on choice of algebra.

THEOREM 3.2.1. *The differential equation*

(3.2.1) $$\frac{dy}{dx} = 3x \otimes x$$

in complex field has form of system of differential equations

(3.2.2) $$\begin{cases} \dfrac{\partial y^0}{\partial x^0} = 3(x^0)^2 - 3(x^1)^2 & \dfrac{\partial y^0}{\partial x^1} = -6x^0 x^1 \\ \dfrac{\partial y^1}{\partial x^0} = 6x^0 x^1 & \dfrac{\partial y^1}{\partial x^1} = 3(x^0)^2 - 3(x^1)^2 \end{cases}$$

with respect to the basis

$$e_{C \cdot 0} = 1 \quad e_{C \cdot 1} = i$$

PROOF. Since product in complex field is commutative, we can write differential equation (3.2.1) as follows

(3.2.3) $$dy = 3x^2 \, dx$$

If we represent differentials dx, dy as vector-column, then, according to the theorem [6]-2.5.1 and lemma 3.1.5, the equation (3.2.3) gets following form

$$(3.2.4) \quad \begin{pmatrix} dy^0 \\ dy^1 \end{pmatrix} = 3 \begin{pmatrix} (x^0)^2 - (x^1)^2 & -2x^0 x^1 \\ 2x^0 x^1 & (x^0)^2 - (x^1)^2 \end{pmatrix} \begin{pmatrix} dx^0 \\ dx^1 \end{pmatrix}$$

Since the matrix of derivative has following form

$$\frac{dy}{dx} = \begin{pmatrix} \dfrac{\partial y^0}{\partial x^0} & \dfrac{\partial y^0}{\partial x^1} \\ \dfrac{\partial y^1}{\partial x^0} & \dfrac{\partial y^1}{\partial x^1} \end{pmatrix}$$

then the equality

$$(3.2.5) \quad \begin{pmatrix} \dfrac{\partial y^0}{\partial x^0} & \dfrac{\partial y^0}{\partial x^1} \\ \dfrac{\partial y^1}{\partial x^0} & \dfrac{\partial y^1}{\partial x^1} \end{pmatrix} = \begin{pmatrix} 3(x^0)^2 - 3(x^1)^2 & -6x^0 x^1 \\ 6x^0 x^1 & 3(x^0)^2 - 3(x^1)^2 \end{pmatrix}$$

follows from (3.2.4). The system of differential equations (3.2.2) follows from the equality (3.2.5). □

LEMMA 3.2.2. *Let*
$$x = x^0 + x^1 i$$
be complex number. Then

$$(3.2.6) \quad x^3 = (x^0)^3 - 3x^0(x^1)^2 + (3(x^0)^2 x^1 - (x^1)^3)i$$

PROOF. According to the lemma 3.1.5,

$$(3.2.7) \quad \begin{aligned} x^3 = x^2 x &= ((x^0)^2 - (x^1)^2 + 2x^0 x^1 i)(x^0 + x^1 i) \\ &= ((x^0)^2 - (x^1)^2)x^0 + 2(x^0)^2 x^1 i + ((x^0)^2 - (x^1)^2)x^1 i - 2x^0(x^1)^2 \\ &= (x^0)^3 - x^0(x^1)^2 + 2(x^0)^2 x^1 i + ((x^0)^2 x^1 - (x^1)^3)i - 2x^0(x^1)^2 \end{aligned}$$

The equality (3.2.6) follows from the equality (3.2.7) □

THEOREM 3.2.3. *The map*
$$y = x^3 + C$$
is solution of the system of differential equations (3.2.2) *in complex field with initial condition*
$$x = 0 \quad y^0 = C^0 \quad y^1 = C^1 \quad C = C^0 + C^1 i$$

PROOF. The map

$$(3.2.8) \quad y^0 = (x^0)^3 - 3x^0(x^1)^2 + C_1^0(x^1)$$

is solution of the differential equation

$$\frac{\partial y^0}{\partial x^0} = 3(x^0)^2 - 3(x^1)^2$$

From (3.2.2), (3.2.8), it follows that

$$\frac{\partial y^0}{\partial x^1} = -6x^0 x^1 + \frac{dC_1^0}{dx^1} = -6x^0 x^1 \tag{3.2.9}$$

The map

$$C_1^0(x^1) = C^0 \tag{3.2.10}$$

is solution of the differential equation (3.2.9). The equality

$$y^0 = (x^0)^3 - 3x^0(x^1)^2 + C^0 \tag{3.2.11}$$

follows from equalities (3.2.8), (3.2.10).

The map

$$y^1 = 3(x^0)^2 x^1 + C_1^1(x^1) \tag{3.2.12}$$

is solution of the differential equation

$$\frac{\partial y^1}{\partial x^0} = 6x^0 x^1$$

From (3.2.2), (3.2.12), it follows that

$$\frac{\partial y^1}{\partial x^1} = 3(x^0)^2 + \frac{dC_1^1}{dx^1} = 3(x^0)^2 - 3(x^1)^2 \tag{3.2.13}$$

The map

$$C_1^1(x^1) = -(x^1)^3 + C^1 \tag{3.2.14}$$

is solution of the differential equation (3.2.13). The equality

$$y^1 = 3(x^0)^2 x^1 - (x^1)^3 + C^1 \tag{3.2.15}$$

follows from equalities (3.2.12), (3.2.14).

The theorem follows from the comparison of equality (3.2.6) and equalities (3.2.11), (3.2.15). □

THEOREM 3.2.4. *The differential equation*

$$\frac{dy}{dx} = 3x \otimes x \tag{3.2.16}$$

in non commutative algebra does not possess a solution.

PROOF. We consider **method of successive differentiation** to solve differential equation (3.2.16). Differentiating one after another equation (3.2.16), we get the chain of equations

$$\frac{d^2 y}{dx^2} = 3 \otimes 1 \otimes x + 3x \otimes 1 \otimes 1 \tag{3.2.17}$$

$$\frac{d^3 y}{dx^3} = 6 \otimes 1 \otimes 1 \otimes 1 \tag{3.2.18}$$

The expansion into Taylor series

$$y = x^3 + C$$

follows from equations (3.2.16), (3.2.17), (3.2.18) and from initial condition. According to the theorem A.1.9

$$\frac{dx^3}{dx} \neq 3x \otimes x \tag{3.2.19}$$

The theorem follows from the statement (3.2.19). $\square$

LEMMA 3.2.5. *In quaternion algebra*
$J_l(a)J_r(b) = J_r(b)J_l(a)$

$$\tag{3.2.20}
= \begin{pmatrix}
a^0b^0 - a^1b^1 & -a^0b^1 - a^1b^0 & -a^0b^2 - a^1b^3 & -a^0b^3 + a^1b^2 \\
-a^2b^2 - a^3b^3 & +a^2b^3 - a^3b^2 & -a^2b^0 + a^3b^1 & -a^2b^1 - a^3b^0 \\
\hline
a^1b^0 + a^0b^1 & -a^1b^1 + a^0b^0 & -a^1b^2 + a^0b^3 & -a^1b^3 - a^0b^2 \\
-a^3b^2 + a^2b^3 & +a^3b^3 + a^2b^2 & -a^3b^0 - a^2b^1 & -a^3b^1 + a^2b^0 \\
\hline
a^2b^0 + a^3b^1 & -a^2b^1 + a^3b^0 & -a^2b^2 + a^3b^3 & -a^2b^3 - a^3b^2 \\
+a^0b^2 - a^1b^3 & -a^0b^3 - a^1b^2 & +a^0b^0 + a^1b^1 & +a^0b^1 - a^1b^0 \\
\hline
a^3b^0 - a^2b^1 & -a^3b^1 - a^2b^0 & -a^3b^2 - a^2b^3 & -a^3b^3 + a^2b^2 \\
+a^1b^2 + a^0b^3 & -a^1b^3 + a^0b^2 & +a^1b^0 - a^0b^1 & +a^1b^1 + a^0b^0
\end{pmatrix}$$

PROOF. The product of matrices $J_l(a)$, $J_r(b)$ has form

$$\begin{pmatrix}
a^0 & -a^1 & -a^2 & -a^3 \\
a^1 & a^0 & -a^3 & a^2 \\
a^2 & a^3 & a^0 & -a^1 \\
a^3 & -a^2 & a^1 & a^0
\end{pmatrix}
\begin{pmatrix}
b^0 & -b^1 & -b^2 & -b^3 \\
b^1 & b^0 & b^3 & -b^2 \\
b^2 & -b^3 & b^0 & b^1 \\
b^3 & b^2 & -b^1 & b^0
\end{pmatrix}$$

$$\tag{3.2.21}
= \begin{pmatrix}
a^0b^0 - a^1b^1 & -a^0b^1 - a^1b^0 & -a^0b^2 - a^1b^3 & -a^0b^3 + a^1b^2 \\
-a^2b^2 - a^3b^3 & +a^2b^3 - a^3b^2 & -a^2b^0 + a^3b^1 & -a^2b^1 - a^3b^0 \\
\hline
a^1b^0 + a^0b^1 & -a^1b^1 + a^0b^0 & -a^1b^2 + a^0b^3 & -a^1b^3 - a^0b^2 \\
-a^3b^2 + a^2b^3 & +a^3b^3 + a^2b^2 & -a^3b^0 - a^2b^1 & -a^3b^1 + a^2b^0 \\
\hline
a^2b^0 + a^3b^1 & -a^2b^1 + a^3b^0 & -a^2b^2 + a^3b^3 & -a^2b^3 - a^3b^2 \\
+a^0b^2 - a^1b^3 & -a^0b^3 - a^1b^2 & +a^0b^0 + a^1b^1 & +a^0b^1 - a^1b^0 \\
\hline
a^3b^0 - a^2b^1 & -a^3b^1 - a^2b^0 & -a^3b^2 - a^2b^3 & -a^3b^3 + a^2b^2 \\
+a^1b^2 + a^0b^3 & -a^1b^3 + a^0b^2 & +a^1b^0 - a^0b^1 & +a^1b^1 + a^0b^0
\end{pmatrix}$$

The product of matrices $J_r(b)$, $J_l(a)$ has form

$$\begin{pmatrix} b^0 & -b^1 & -b^2 & -b^3 \\ b^1 & b^0 & b^3 & -b^2 \\ b^2 & -b^3 & b^0 & b^1 \\ b^3 & b^2 & -b^1 & b^0 \end{pmatrix} \begin{pmatrix} a^0 & -a^1 & -a^2 & -a^3 \\ a^1 & a^0 & -a^3 & a^2 \\ a^2 & a^3 & a^0 & -a^1 \\ a^3 & -a^2 & a^1 & a^0 \end{pmatrix}$$

(3.2.22)
$$= \begin{pmatrix} \begin{array}{c} a^0b^0 - a^1b^1 \\ -a^2b^2 - a^3b^3 \end{array} & \begin{array}{c} -a^0b^1 - a^1b^0 \\ +a^2b^3 - a^3b^2 \end{array} & \begin{array}{c} -a^0b^2 - a^1b^3 \\ -a^2b^0 + a^3b^1 \end{array} & \begin{array}{c} -a^0b^3 + a^1b^2 \\ -a^2b^1 - a^3b^0 \end{array} \\ \hline \begin{array}{c} a^1b^0 + a^0b^1 \\ -a^3b^2 + a^2b^3 \end{array} & \begin{array}{c} -a^1b^1 + a^0b^0 \\ +a^3b^3 + a^2b^2 \end{array} & \begin{array}{c} -a^1b^2 + a^0b^3 \\ -a^3b^0 - a^2b^1 \end{array} & \begin{array}{c} -a^1b^3 - a^0b^2 \\ -a^3b^1 + a^2b^0 \end{array} \\ \hline \begin{array}{c} a^2b^0 + a^3b^1 \\ +a^0b^2 - a^1b^3 \end{array} & \begin{array}{c} -a^2b^1 + a^3b^0 \\ -a^0b^3 - a^1b^2 \end{array} & \begin{array}{c} -a^2b^2 + a^3b^3 \\ +a^0b^0 + a^1b^1 \end{array} & \begin{array}{c} -a^2b^3 - a^3b^2 \\ +a^0b^1 - a^1b^0 \end{array} \\ \hline \begin{array}{c} a^3b^0 - a^2b^1 \\ +a^1b^2 + a^0b^3 \end{array} & \begin{array}{c} -a^3b^1 - a^2b^0 \\ -a^1b^3 + a^0b^2 \end{array} & \begin{array}{c} -a^3b^2 - a^2b^3 \\ +a^1b^0 - a^0b^1 \end{array} & \begin{array}{c} -a^3b^3 + a^2b^2 \\ +a^1b^1 + a^0b^0 \end{array} \end{pmatrix}$$

The equality (3.2.20) follows from equalities (3.2.21), (3.2.22). $\square$

LEMMA 3.2.6. *The product of matrices* $J_r(a)$, $J_l(a)$ *has form*

(3.2.23)
$$\begin{pmatrix} \begin{array}{c} a^0a^0 - a^1a^1 \\ -a^2a^2 - a^3a^3 \end{array} & -2a^0a^1 & -2a^0a^2 & -2a^0a^3 \\ \hline 2a^0a^1 & \begin{array}{c} -a^1a^1 + a^0a^0 \\ +a^3a^3 + a^2a^2 \end{array} & -2a^1a^2 & -2a^1a^3 \\ \hline 2a^2a^0 & -2a^2a^1 & \begin{array}{c} -a^2a^2 + a^3a^3 \\ +a^0a^0 + a^1a^1 \end{array} & -2a^2a^3 \\ \hline 2a^3a^0 & -2a^3a^1 & -2a^2a^3 & \begin{array}{c} -a^3a^3 + a^2a^2 \\ +a^1a^1 + a^0a^0 \end{array} \end{pmatrix}$$

PROOF. According to the lemma 3.2.5, the product of matrices $J_r(a)$, $J_l(a)$ has form

(3.2.24)
$$\begin{pmatrix} a^0a^0 - a^1a^1 & -a^0a^1 - a^1a^0 & -a^0a^2 - a^1a^3 & -a^0a^3 + a^1a^2 \\ -a^2a^2 - a^3a^3 & +a^2a^3 - a^3a^2 & -a^2a^0 + a^3a^1 & -a^2a^1 - a^3a^0 \\ \hline a^1a^0 + a^0a^1 & -a^1a^1 + a^0a^0 & -a^1a^2 + a^0a^3 & -a^1a^3 - a^0a^2 \\ -a^3a^2 + a^2a^3 & +a^3a^3 + a^2a^2 & -a^3a^0 - a^2a^1 & -a^3a^1 + a^2a^0 \\ \hline a^2a^0 + a^3a^1 & -a^2a^1 + a^3a^0 & -a^2a^2 + a^3a^3 & -a^2a^3 - a^3a^2 \\ +a^0a^2 - a^1a^3 & -a^0a^3 - a^1a^2 & +a^0a^0 + a^1a^1 & +a^0a^1 - a^1a^0 \\ \hline a^3a^0 - a^2a^1 & -a^3a^1 - a^2a^0 & -a^3a^2 - a^2a^3 & -a^3a^3 + a^2a^2 \\ +a^1a^2 + a^0a^3 & -a^1a^3 + a^0a^2 & +a^1a^0 - a^0a^1 & +a^1a^1 + a^0a^0 \end{pmatrix}$$

The equality (3.2.23) follows from the equality (3.2.24). □

THEOREM 3.2.7. *The differential equation* (3.2.16) *in quaternion algebra has form of system of differential equations*

(3.2.25)
$$\begin{cases} \dfrac{\partial y^0}{\partial x^0} = 3(x^0)^2 - 3(x^1)^2 - 3(x^2)^2 - 3(x^3)^2 \\ \dfrac{\partial y^0}{\partial x^1} = -6x^0 x^1 \\ \dfrac{\partial y^0}{\partial x^2} = -6x^0 x^2 \\ \dfrac{\partial y^0}{\partial x^3} = -6x^0 x^3 \end{cases}$$

(3.2.26)
$$\begin{cases} \dfrac{\partial y^1}{\partial x^0} = 6x^0 x^1 \\ \dfrac{\partial y^1}{\partial x^1} = 3(x^0)^2 - 3(x^1)^2 + 3(x^2)^2 + 3(x^3)^2 \\ \dfrac{\partial y^1}{\partial x^2} = -6x^1 x^2 \\ \dfrac{\partial y^1}{\partial x^3} = -6x^1 x^3 \end{cases}$$

$$(3.2.27) \begin{cases} \dfrac{\partial y^2}{\partial x^0} = 6x^0 x^2 \\ \dfrac{\partial y^2}{\partial x^1} = -6x^1 x^2 \\ \dfrac{\partial y^2}{\partial x^2} = -3(x^2)^2 + 3(x^3)^2 + 3(x^0)^2 + 3(x^1)^2 \\ \dfrac{\partial y^2}{\partial x^3} = -6x^2 x^3 \end{cases}$$

$$(3.2.28) \begin{cases} \dfrac{\partial y^3}{\partial x^0} = 6x^0 x^3 \\ \dfrac{\partial y^3}{\partial x^1} = -6x^1 x^3 \\ \dfrac{\partial y^3}{\partial x^2} = -6x^2 x^3 \\ \dfrac{\partial y^3}{\partial x^3} = -3(x^3)^2 + 3(x^2)^2 + 3(x^1)^2 + 3(x^0)^2 \end{cases}$$

with respect to the basis

$$e_0 = 1 \quad e_1 = i \quad e_2 = j \quad e_3 = k$$

PROOF. We can write differential equation (3.2.16) as follows

$$(3.2.29) \qquad dy = 3x\, dx\, x$$

If we represent differentials dx, dy as vector-column, then, according to the lemma 3.2.5, the equation (3.2.29) gets following form

$$(3.2.30) \qquad \begin{pmatrix} dy^0 \\ dy^1 \\ dy^2 \\ dy^3 \end{pmatrix} = 3 J_l(x) J_r(x) \begin{pmatrix} dx^0 \\ dx^1 \\ dx^2 \\ dx^3 \end{pmatrix}$$

Since the matrix of derivative has following form

$$\frac{dy}{dx} = \begin{pmatrix} \dfrac{\partial y^0}{\partial x^0} & \dfrac{\partial y^0}{\partial x^1} & \dfrac{\partial y^0}{\partial x^2} & \dfrac{\partial y^0}{\partial x^3} \\ \dfrac{\partial y^1}{\partial x^0} & \dfrac{\partial y^1}{\partial x^1} & \dfrac{\partial y^1}{\partial x^2} & \dfrac{\partial y^1}{\partial x^3} \\ \dfrac{\partial y^2}{\partial x^0} & \dfrac{\partial y^2}{\partial x^1} & \dfrac{\partial y^2}{\partial x^2} & \dfrac{\partial y^2}{\partial x^3} \\ \dfrac{\partial y^3}{\partial x^0} & \dfrac{\partial y^3}{\partial x^1} & \dfrac{\partial y^3}{\partial x^2} & \dfrac{\partial y^3}{\partial x^3} \end{pmatrix}$$

then the equality

$$\frac{1}{3}\begin{pmatrix} \frac{\partial y^0}{\partial x^0} & \frac{\partial y^0}{\partial x^1} & \frac{\partial y^0}{\partial x^2} & \frac{\partial y^0}{\partial x^3} \\ \frac{\partial y^1}{\partial x^0} & \frac{\partial y^1}{\partial x^1} & \frac{\partial y^1}{\partial x^2} & \frac{\partial y^1}{\partial x^3} \\ \frac{\partial y^2}{\partial x^0} & \frac{\partial y^2}{\partial x^1} & \frac{\partial y^2}{\partial x^2} & \frac{\partial y^2}{\partial x^3} \\ \frac{\partial y^3}{\partial x^0} & \frac{\partial y^3}{\partial x^1} & \frac{\partial y^3}{\partial x^2} & \frac{\partial y^3}{\partial x^3} \end{pmatrix} =$$

(3.2.31)
$$\begin{pmatrix} x^0x^0 - x^1x^1 & -2x^0x^1 & -2x^0x^2 & -2x^0x^3 \\ -x^2x^2 - x^3x^3 & & & \\ 2x^0x^1 & -x^1x^1 + x^0x^0 & -2x^1x^2 & -2x^1x^3 \\ & +x^3x^3 + x^2x^2 & & \\ 2x^2x^0 & -2x^2x^1 & -x^2x^2 + x^3x^3 & -2x^2x^3 \\ & & +x^0x^0 + x^1x^1 & \\ 2x^3x^0 & -2x^3x^1 & -2x^2x^3 & -x^3x^3 + x^2x^2 \\ & & & +x^1x^1 + x^0x^0 \end{pmatrix}$$

follows from (3.2.30). The system of differential equations (3.2.25), (3.2.26), (3.2.27), (3.2.28) follows from the equality (3.2.31). □

THEOREM 3.2.8. *The system of differential equations* (3.2.25), (3.2.26), (3.2.27), (3.2.28) *does not possess a solution.*

PROOF. The map
$$y^1 = 3(x^0)^2 x^1 + C_1^1(x^1, x^2, x^3) \tag{3.2.32}$$
is solution of the differential equation
$$\frac{\partial y^1}{\partial x^0} = 6x^0 x^1$$
From (3.2.26), (3.2.32), it follows that
$$\frac{\partial y^1}{\partial x^1} = 3(x^0)^2 + \frac{\partial C_1^1}{\partial x^1} = 3(x^0)^2 - 3(x^1)^2 + 3(x^2)^2 + 3(x^3)^2 \tag{3.2.33}$$
The map
$$C_1^1(x^1, x^2, x^3) = -(x^1)^3 + 3(x^2)^2 x^1 + 3(x^3)^2 x^1 + C_2^1(x^2, x^3) \tag{3.2.34}$$
is solution of the differential equation (3.2.33). The equality
$$y^1 = 3(x^0)^2 x^1 - (x^1)^3 + 3(x^2)^2 x^1 + 3(x^3)^2 x^1 + C_2^1(x^2, x^3) \tag{3.2.35}$$

follows from equalities (3.2.32), (3.2.34). From (3.2.26), (3.2.35), it follows that

$$\frac{\partial y^1}{\partial x^2} = 6x^2 x^1 + \frac{\partial C_2^1}{\partial x^2} = -6x^1 x^2 \tag{3.2.36}$$

From the equation (3.2.36), it follows that the map C_2^1 depends on x^1; this contradicts to the statement that the map C_2^1 does not depend on x^1. Therefore, the system of differential equations (3.2.25), (3.2.26), (3.2.27), (3.2.28) does not possess a solution. □

3.3. Before Going Any Further

Before going any further, we ask questions that need to be answered.

QUESTION 3.3.1. *If D-algebra B has finite basis $\bar{\bar{e}}$, then we can write the differential equation*

$$\frac{df(x)}{dx} = g(x) \tag{3.3.1}$$

as system of differential equations

$$\frac{\partial y^i}{\partial x^j} = g_j^i \quad y = y^i e_{B \cdot i} \quad x = x^i e_{A \cdot i} \tag{3.3.2}$$

What is the relationship between the differential equation (3.3.1) and system of differential equations (3.3.2)? □

QUESTION 3.3.2. *What is the condition for the integrability of the differential equation (3.3.1)?* □

QUESTION 3.3.3. *Can we write down a differential equation (3.3.1) given the system of differential equations (3.3.2)?* □

QUESTION 3.3.4. *Systems of differential equations (3.1.2), (3.1.15), (3.2.2) are completely integrable.*[3.3] *The system of differential equations (3.2.25), (3.2.26), (3.2.27), (3.2.28) is not completely integrable. Is the requirement complete integrability of systems of differential equations (3.3.2) equivalent to the requirement of integrability of corresponding differential equation (3.3.1)?* □

3.4. Forms of representation of differential equations

The theorem 3.4.1 answers the question 3.3.1.

THEOREM 3.4.1. *Let B be free finite dimensional associative D-algebra. Let $\bar{\bar{e}}_A$ be basis of D module A. Let $\bar{\bar{e}}_B$ be basis of D module B. Let $\overline{\overline{F}}$ be the basis of left $B \otimes B$-module $\mathcal{L}(D; B \to B)$ and*

$$G : A \to B$$

[3.3] See the definition of completely integrable system of differential equations on page 2 in [14].

be linear map of maximal rank such that $\ker G \subseteq \ker g$. Let $g^{k \cdot ij}$ be standard components of the map g. Let C^p_{kl} be structural constants of algebra B. Then we can write the differential equation (3.3.1) as system of differential equations

$$\frac{\partial y^k}{\partial x^l} = g^{k \cdot ij} F_{k \cdot r}{}^m G^r_l C^p_{im} C^k_{pj} \tag{3.4.1}$$

PROOF. The theorem follows from the theorem 2.4.15. □

Based on the theorem 3.4.1, we consider the question 3.3.3. However, the answer to this question is not straightforward. We begin with the theorem 3.4.2.

THEOREM 3.4.2. *Let B be free finite dimensional associative D-algebra. Let $\overline{\overline{e}}$ be basis of D module B. Let $\overline{\overline{F}}$ be the basis of left $B \otimes B$-module $\mathcal{L}(D; B \to B)$ and*

$$G : A \to B$$

be linear map of maximal rank such that $\ker G \subseteq \ker g$. Let C^p_{kl} be structural constants of algebra B. Consider matrix

$$\mathcal{C} = \left(\mathcal{C}^{\cdot k}_{m \cdot ij}\right) = \left(C^p_{im} C^k_{pj}\right)$$

whose rows and columns are indexed by ${}^{\cdot k}_m$ and ${}_{\cdot ij}$, respectively. If matrix $\mathcal{C}$ is nonsingular, then we can write the system of differential equations

$$\frac{\partial y^i}{\partial x^j} = g^i_j \quad y = y^i e_{B \cdot i} \quad x = x^i e_{A \cdot i}$$

as differential equation

$$\frac{dy}{dx} = g^{k \cdot ij} (e_{B \cdot i} \otimes e_{B \cdot j}) \circ F_k \circ G \tag{3.4.2}$$

where standard components $g^{k \cdot ij}$ of map g are solution of system of linear equations

$$g^k_l = g^{k \cdot ij} F_{k \cdot r}{}^m G^r_l C_{B \cdot im}{}^p C_{B \cdot pj}{}^k$$

If matrix $\mathcal{C}$ is singular, then we can write the differential equation (3.4.2), if condition

$$\mathrm{rank}\left(\mathcal{C}^{\cdot k}_{m \cdot ij} \quad g^k_m\right) = \mathrm{rank}\,\mathcal{C}$$

is satisfied.

PROOF. The theorem follows from the theorem 2.4.15. □

Analysis of the theorem 3.4.2 shows that it is not sufficient to write the equation (3.4.2) to answer the question 3.3.3. The reason for this is that, according to the construction, maps $g^{k \cdot ij}$ depend on coordinates of A-number x. However, it should be noted that the map

$$x = x^i e_i \to x^j$$

is the linear map. So, if D-module A is D-algebra, then there exists effective procedure to answer the question 3.3.3.

REMARK 3.4.3. *The statement of this remark is prelimenary, because I did not finished research in this direction. Standard components of the map g is universal form of representation of the map, but not the only and not the most expressive. Let $\overline{\overline{E}}$ be basis of B-module $\mathcal{L}(D; B \to B)$. Then the map g has representation*

$$g(x) = g^k(x) \circ E_k \quad g^k : A \to B$$

□

3.5. Condition of Integrability

To answer the question 3.3.2, I recall that the map g is differential form (the definition [8]-7.3.2).

THEOREM 3.5.1. *The differential equation*

$$\frac{df(x)}{dx} = g(x)$$

is integrable iff

$$dg = 0$$

PROOF. The theorem follows from the theorem [8]-8.3.1. □

Consider first how the theorem 3.5.1 works in case of differential equations (3.1.1), (3.2.16).

EXAMPLE 3.5.2. *Consider the differential form*

$$g(x) = x \otimes 1 + 1 \otimes x$$

According to theorems A.1.1, A.1.2, A.1.4, A.1.15,

$$\frac{dg}{dx} = 1 \otimes_2 1 \otimes_1 1 + 1 \otimes_1 1 \otimes_2 1$$

Index accompanying symbol $\otimes$, shows which argument of polylinear map should be written instead of corresponding symbol $\otimes$. For instance

(3.5.1) $$\frac{dg}{dx} \circ (h_1, h_2) = (1 \otimes_2 1 \otimes_1 1 + 1 \otimes_1 1 \otimes_2 1) \circ (h_1, h_2) = h_2 h_1 + h_1 h_2$$

From the equality (3.5.1), it follows that bilinear map $\frac{dg}{dx}$ is symmetrical. According to the definition [8]-7.4.1

$$dg = 0$$

According to the theorem 3.5.1, the differential equation (3.1.1) is integrable (the theorem 3.1.7). □

EXAMPLE 3.5.3. *Consider the differential form*

$$g(x) = x \otimes x$$

According to theorems A.1.2, A.1.15,

$$\frac{dg}{dx} = 1 \otimes_2 1 \otimes_1 x + x \otimes_1 1 \otimes_2 1$$

3.5. Condition of Integrability

Index accompanying symbol $\otimes$, shows which argument of polylinear map should be written instead of corresponding symbol $\otimes$. For instance

(3.5.2) $\quad \dfrac{dg}{dx} \circ (h_1, h_2) = (1 \otimes_2 1 \otimes_1 x + x \otimes_1 1 \otimes_2 1) \circ (h_1, h_2) = h_2 h_1 x + x h_1 h_2$

From the equality (3.5.2) and from the definition [8]-7.4.1, it follows that

(3.5.3) $\quad\quad\quad\quad\quad dg = (h_2 h_1 - h_1 h_2) x + x (h_1 h_2 - h_2 h_1)$

From the equality (3.5.3), it follows that

$$dg \ne 0$$

According to the theorem 3.5.1, the differential equation (3.2.16). is not integrable (the theorem 3.2.4). □

The theorem 3.5.4 answers the question 3.3.2.

THEOREM 3.5.4. *Let B be free finite dimensional associative D-algebra. Let $\overline{\overline{e}}$ be basis of D module B. Let $\overline{\overline{F}}$ be the basis of left $B \otimes B$-module $\mathcal{L}(D; B \to B)$ and G be the linear map*

$$G : A \to B$$

of maximal rank such that $\ker G \subseteq \ker g$. Let $g^{k \cdot ij}$ be standard components of the map

$$g : A \to \mathcal{L}(D; A \to B)$$

The differential equation

$$\frac{dy}{dx} = g^{k \cdot ij} (e_{B \cdot i} \otimes e_{B \cdot j}) \circ F_k \circ G$$

is integrable iff

(3.5.4) $\quad \begin{aligned} & \left(\dfrac{dg^{k \cdot ij}(x)}{dx} \circ h_2 \right) (e_{B \cdot i}(F_k \circ G \circ h_1) e_{B \cdot j}) \\ & = \left(\dfrac{dg^{k \cdot ij}(x)}{dx} \circ h_1 \right) (e_{B \cdot i}(F_k \circ G \circ h_2) e_{B \cdot j}) \end{aligned}$

PROOF. Since

(3.5.5) $\quad g \circ h_1 = g^{k \cdot ij}(e_i \otimes e_j) \circ F_k \circ G \circ h_1 = g^{k \cdot ij}(e_i(F_k \circ G \circ h_1) e_j)$

then, according to the theorem [8]-8.2.3,

(3.5.6) $\quad \dfrac{dg}{dx} \circ (h_1, h_2) = \left(\dfrac{dg^{k \cdot ij}}{dx} \circ h_2 \right) (e_i(F_k \circ G \circ h_1) e_j)$

The equality (3.5.4) follows from the equality (3.5.6), from the theorem 3.5.1 and the definition [8]-7.4.1. □

The theorem 3.5.5 answers the question 3.3.4.

THEOREM 3.5.5. *Let B be free finite dimensional associative D-algebra. Let $\bar{\bar{e}}_A$ be basis of D module A. Let $\bar{\bar{e}}_B$ be basis of D module B. The differential equation*

$$\frac{dy}{dx} = g(x) \tag{3.5.7}$$

is integrable iff corresponding system of differential equations

$$\frac{\partial y^i}{\partial x^j} = g^i_j \quad y = y^i e_{B \cdot i} \quad x = x^i e_{A \cdot i} \tag{3.5.8}$$

is completely integrable.

PROOF. Let $\bar{\bar{F}}$ be the basis of left $B \otimes B$-module $\mathcal{L}(D; B \to B)$ and G be the linear map

$$G : A \to B$$

of maximal rank such that $\ker G \subseteq \ker .L$ et $g^{k \cdot ij}$ be standard components of the map

$$g : A \to \mathcal{L}(D; A \to B)$$

Then the differential equation (3.5.7) has form

$$\frac{dy}{dx} = g^{k \cdot ij}(e_{B \cdot i} \otimes e_{B \cdot j}) \circ F_k \circ G \tag{3.5.9}$$

According to the theorem 3.5.4, the differential equation (3.5.7) is integrable iff

$$\begin{aligned}
&\left(\frac{dg^{k \cdot ij}(x)}{dx} \circ h_2\right)(e_{B \cdot i}(F_k \circ G \circ h_1)e_{B \cdot j}) \\
&= \left(\frac{dg^{k \cdot ij}(x)}{dx} \circ h_1\right)(e_{B \cdot i}(F_k \circ G \circ h_2)e_{B \cdot j})
\end{aligned} \tag{3.5.10}$$

Let

$$h_1 = h_1^k e_k \quad h_2 = h_2^l e_l$$

Then we can write the equality (3.5.10) as

$$\begin{aligned}
&\left(\frac{dg^{k \cdot ij}(x)}{dx} \circ (h_2^l e_{1 \cdot l})\right)(e_{2 \cdot i}(F_k \circ G \circ (h_1^k e_{1 \cdot k}))e_{2 \cdot j}) \\
&= \left(\frac{dg^{k \cdot ij}(x)}{dx} \circ (h_1^k e_{1 \cdot k})\right)(e_{2 \cdot i}(F_k \circ G \circ (h_2^l e_{1 \cdot l}))e_{2 \cdot j})
\end{aligned} \tag{3.5.11}$$

Since h_1, h_2 are arbitrary A-numbers, then the equality

$$\begin{aligned}
&\left(\frac{dg^{k \cdot ij}(x)}{dx} \circ e_{1 \cdot l}\right)(e_{2 \cdot i}(F_k \circ G \circ e_{1 \cdot k})e_{2 \cdot j}) \\
&= \left(\frac{dg^{k \cdot ij}(x)}{dx} \circ e_{1 \cdot k}\right)(e_{2 \cdot i}(F_k \circ G \circ e_{1 \cdot l})e_{2 \cdot j})
\end{aligned} \tag{3.5.12}$$

follows from the equality (3.5.11). Since

$$F_k \circ G \circ e_{1 \cdot k} = F_{kr}^{\ m} G_k^r e_{2 \cdot m}$$

3.5. Condition of Integrability

then the equality

$$\begin{aligned}(3.5.13) \quad & \left(\frac{dg^{k\cdot ij}(x)}{dx}\circ e_{1\cdot l}\right)F_{kr}^{\ m}G_{k}^{r}e_{2\cdot i}e_{2\cdot m}e_{2\cdot j} \\ & = \left(\frac{dg^{k\cdot ij}(x)}{dx}\circ e_{1\cdot k}\right)F_{kr}^{\ m}G_{l}^{r}e_{2\cdot i}e_{2\cdot m}e_{2\cdot j}\end{aligned}$$

follows from the equality (3.5.12). The equality

$$\begin{aligned}(3.5.14) \quad & \left(\frac{dg^{ij}(x)}{dx}\circ e_{1\cdot l}\right)F_{kr}^{\ m}G_{k}^{r}C_{2\cdot im}^{\ \ p}C_{2\cdot pj}^{\ \ r}e_{2\cdot r} \\ & = \left(\frac{dg^{k\cdot ij}(x)}{dx}\circ e_{1\cdot k}\right)F_{kr}^{\ m}G_{l}^{r}C_{2\cdot im}^{\ \ p}C_{2\cdot pj}^{\ \ r}e_{2\cdot r}\end{aligned}$$

follows from the equality (3.5.13). The equality

$$(3.5.15) \quad \left(\frac{dg_{k}^{r}(x)}{dx}\circ e_{1\cdot l}\right)e_{2\cdot r} = \left(\frac{dg_{l}^{r}(x)}{dx}\circ e_{1\cdot k}\right)e_{2\cdot r}$$

follows from the equality (3.5.14). The equality

$$(3.5.16) \quad \frac{dg_{k}^{r}(x)}{dx}\circ e_{1\cdot l} = \frac{dg_{l}^{r}(x)}{dx}\circ e_{1\cdot k}$$

follows from the equality (3.5.15). $\square$

CHAPTER 4

Differential Equation of First Order

4.1. Differential Equation with Separated Variables

In the theory of differential equations over field, the first differential equation which we study is separable equation

$$\frac{dy}{dx} = \frac{M(x)}{N(y)} \tag{4.1.1}$$

As soon as we separate variables, we can write differential equation (4.1.1) as

$$N(y)dy = M(x)dx \tag{4.1.2}$$

It is easy to integrate differential equation (4.1.2).

In noncommutative D-algebra, dx and dy belong in general to different D-modules. At the same time, a solution of the differential equation (4.1.2) is implicit function of variables x and y. So I formulate the problem as follows.

Let X, Y be Banach D-modules. Let A be Banach D-algebra. Consider maps

$$f : X \to \mathcal{L}(D; X \to A)$$
$$g : Y \to \mathcal{L}(D; Y \to A)$$

Differential equation

$$\frac{dy}{dx} = g(y) \circ f(x) \tag{4.1.3}$$

is called **differential separable equation**.

In noncommutative D-algebra, even this is division algebra, operation of separation of variables may be impossible. However, if there exists the map

$$h : Y \to \mathcal{L}(D; Y \to A)$$

such that

$$h \circ g = 1 \otimes 1$$

then we can write the differential equation (4.1.3) in the following form

$$h(y) \circ dy = f(x) \circ dx$$

Let X, Y be Banach D-modules. Let A be Banach D-algebra. Consider maps

$$M : X \to \mathcal{L}(D; X \to A)$$
$$N : Y \to \mathcal{L}(D; Y \to A)$$

Differential equation

$$M(x) \circ dx + N(y) \circ dy = 0$$

is called **differential equation with separated variables**.

THEOREM 4.1.1. *Let maps*
$$M : X \to \mathcal{L}(D; X \to A)$$
$$N : Y \to \mathcal{L}(D; Y \to A)$$
be integrable. The solution of differential equation

(4.1.4) $$M(x) \circ dx + N(y) \circ dy = 0$$

is implicit function[4.1]

(4.1.5) $$\int M(x) \circ dx + \int N(y) \circ dy = C$$

PROOF. The equality

(4.1.6) $$N(y) \circ dx = N(y) \circ \frac{dy}{dx} \circ dx$$

follows from the definition [8]-3.3.2. Integrability of the differential form in right side of the equality (4.1.6) follows from integrability of the differential form in left side of the equality (4.1.6). According to the theorem A.2.4 the differential form

(4.1.7) $$M(x) \circ dx + N(y) \circ dx = \left(M(x) + N(y) \circ \frac{dy}{dx}\right) \circ dx$$

is integrable. The equality

(4.1.8) $$M(x) + N(y) \circ \frac{dy}{dx} = 0 \otimes 0$$

follows from equalities (4.1.4), (4.1.7). The equality (4.1.5) follows from the equality (4.1.8) and from the theorem A.2.3. $\square$

EXAMPLE 4.1.2. *Consider the differential equation*

(4.1.9) $$(1 \otimes x + x \otimes 1) \circ dx + (1 \otimes y + y \otimes 1) \circ dy = 0$$

According to theorems 4.1.1, A.2.6, implicit function
$$x^2 + y^2 = C$$
is solution of differential equation (4.1.9). $\square$

THEOREM 4.1.3. *The solution of the differential equation*

(4.1.10) $$M(x) \circ dx + N(y) \circ dy = 0$$

with initial condition
$$x_0 = 0 \quad y_0 = C$$

[4.1] The proof of the theorem is based on the proofs on the page [3]-44 and on the page [4]-24; however the proof is different since D-algebra A in general is noncommutative.

has form[4.2]

(4.1.11) $$\int_{x_0}^{x} M(x) \circ dx + \int_{y_0}^{y} N(y) \circ dy = 0$$

PROOF. According to the theorem 4.1.1, implicit function

(4.1.12) $$\int M(x) \circ dx + \int N(y) \circ dy = C$$

is solution of the differential equation (4.1.10). According to the definition 3.1.1, there exist maps

$$P : X \to A$$
$$R : Y \to A$$

such that

(4.1.13) $$P(x) = \int M(x) \circ dx$$

(4.1.14) $$R(y) = \int N(y) \circ dy$$

According to the theorem [8]-6.2.1 and to the definition [8]-8.3.7, equalities

(4.1.15) $$\int M(x) \circ dx = \int_{x_0}^{x} M(x) \circ dx + P(x_0)$$

(4.1.16) $$\int N(y) \circ dy = \int_{y_0}^{y} N(y) \circ dy + R(y_0)$$

follow from equalities (4.1.13), (4.1.14). The equality

(4.1.17) $$\int_{x_0}^{x} M(x) \circ dx + P(x_0) + \int_{y_0}^{y} N(y) \circ dy + R(x_0) = C$$

follows from equalities (4.1.12), (4.1.15), (4.1.16). If $x = x_0$, $y = y_0$, then the equality

(4.1.18) $$P(x_0) + R(x_0) = C$$

follows from the equality (4.1.17). The equality (4.1.11) follows from equalities (4.1.17), (4.1.18). □

[4.2] See also remarks on the page [3]-44 and on the page [4]-24.

4.2. Exact Differential Equation

DEFINITION 4.2.1. *The differential equation*
$$M(x,y) \circ dx + N(x,y) \circ dy = 0$$
where
$$M : (x,y) \in X \times Y \to M(x,y) \in \mathcal{L}(D; X \to A)$$
$$N : (x,y) \in X \times Y \to N(x,y) \in \mathcal{L}(D; Y \to A)$$
is called **exact differential equation**, *if there exists map*
$$u : X \times Y \to A$$
such that

(4.2.1) $$\frac{\partial u(x,y)}{\partial x} = M(x,y)$$

(4.2.2) $$\frac{\partial u(x,y)}{\partial y} = N(x,y)$$

$\square$

THEOREM 4.2.2. *The differential equation*

(4.2.3) $$M(x,y) \circ dx + N(x,y) \circ dy = 0$$

is integrable iff

(4.2.4) $$\frac{\partial M(x,y)}{\partial x} \circ (dx_1, dx_2) = \frac{\partial M(x,y)}{\partial x} \circ (dx_2, dx_1)$$

(4.2.5) $$\frac{\partial N(x,y)}{\partial y} \circ (dy_1, dy_2) = \frac{\partial N(x,y)}{\partial y} \circ (dy_2, dy_1)$$

(4.2.6) $$\frac{\partial M(x,y)}{\partial y} \circ (dx, dy) = \frac{\partial N(x,y)}{\partial x} \circ (dy, dx)$$

PROOF. Let $Z = X \oplus Y$, $z = x \oplus y$. According to definitions [8]-7.3.2, 2.7.5, 4.2.1 and the theorem 2.7.6, the expression

(4.2.7) $$\omega = M(x,y) \circ dx + N(x,y) \circ dy$$

is differential form of degree 1. According to the theorem [8]-8.3.1, differential form ω is integrable iff $d\omega = 0$. Equalities

(4.2.8) $$\begin{aligned}\frac{d\omega}{dz} \circ (z_1, z_2) = & \frac{\partial M(x,y)}{\partial x} \circ (dx_1, dx_2) + \frac{\partial M(x,y)}{\partial y} \circ (dx_1, dy_2) \\ & + \frac{\partial N(x,y)}{\partial x} \circ (dy_1, dx_2) + \frac{\partial N(x,y)}{\partial y} \circ (dy_1, dy_2)\end{aligned}$$

52 4. Differential Equation of First Order

$$\text{(4.2.9)} \quad \frac{d\omega}{dz} \circ (z_2, z_1) = \frac{\partial M(x,y)}{\partial x} \circ (dx_2, dx_1) + \frac{\partial M(x,y)}{\partial y} \circ (dx_2, dy_1)$$
$$+ \frac{\partial N(x,y)}{\partial x} \circ (dy_2, dx_1) + \frac{\partial N(x,y)}{\partial y} \circ (dy_2, dy_1)$$

follow from the equality (4.2.7). The equality

$$\text{(4.2.10)} \quad \begin{aligned} & \frac{\partial M(x,y)}{\partial x} \circ (dx_1, dx_2) + \frac{\partial M(x,y)}{\partial y} \circ (dx_1, dy_2) \\ & + \frac{\partial N(x,y)}{\partial x} \circ (dy_1, dx_2) + \frac{\partial N(x,y)}{\partial y} \circ (dy_1, dy_2) \\ & = \frac{\partial M(x,y)}{\partial x} \circ (dx_2, dx_1) + \frac{\partial M(x,y)}{\partial y} \circ (dx_2, dy_1) \\ & + \frac{\partial N(x,y)}{\partial x} \circ (dy_2, dx_1) + \frac{\partial N(x,y)}{\partial y} \circ (dy_2, dy_1) \end{aligned}$$

follows from equalities (4.2.8), (4.2.9). Equalities (4.2.4), (4.2.5), (4.2.6) follow from the equality (4.2.10). □

Let equalities (4.2.4), (4.2.5), (4.2.6) be true. Our goal is to find the implicit function[4.3] $u(x,y) = C$ which satisfies to equalities (4.2.1), (4.2.2). According to theorems [8]-8.3.1, A.2.1, the equality

$$\text{(4.2.11)} \quad u(x,y) = \int M(x,y) \circ dx + C_1(y)$$

follows from equalities (4.2.1), (4.2.4). The equality

$$\text{(4.2.12)} \quad \frac{\partial}{\partial y} \int M(x,y) \circ dx + \frac{dC_1(y)}{dy} = N(x,y)$$

follows from equalities (4.2.2), (4.2.11). The equality

$$\text{(4.2.13)} \quad \frac{dC_1(y)}{dy} = N(x,y) - \frac{\partial}{\partial y} \int M(x,y) \circ dx$$

follows from the equality (4.2.12). There is variable x in the expression

$$\text{(4.2.14)} \quad N(x,y) - \frac{\partial}{\partial y} \int M(x,y) \circ dx$$

of right side of the equation (4.2.13). In order for the equation (4.2.13) to have a solution, it is necessary that expression (4.2.14) does not depend on x. Consider

---

[4.3] Solving of the equation (4.2.3) is similar to solving of exact differential equation in commutative algebra. See, for instance, the proof of the theorem [3]-2.6.1 as well solving of exact differential equation on pages 33, 34 in the book [4].

derivative of the expression (4.2.14) with respect to x. The equality

(4.2.15)
$$\begin{aligned}
&\frac{\partial}{\partial x}\left(N(x,y)\circ dy - \left(\frac{\partial}{\partial y}\int M(x,y)\circ dx\right)\circ dy\right)\circ dx \\
&= \frac{\partial N(x,y)}{\partial x}\circ (dy,dx) - \left(\frac{\partial^2}{\partial x \partial y}\int M(x,y)\circ dx\right)\circ (dy,dx) \\
&= \frac{\partial N(x,y)}{\partial x}\circ (dy,dx) - \left(\frac{\partial^2}{\partial y \partial x}\int M(x,y)\circ dx\right)\circ (dx,dy) \\
&= \frac{\partial N(x,y)}{\partial x}\circ (dy,dx) - \left(\frac{\partial}{\partial y}\frac{\partial}{\partial x}\int M(x,y)\circ dx\right)\circ dy \\
&= \frac{\partial N(x,y)}{\partial x}\circ (dy,dx) - \frac{\partial M(x,y)}{\partial y}\circ (dx,dy) = 0
\end{aligned}$$

follows from equalities (2.8.9), (4.2.6), (A.2.2).

From theorems [8]-8.3.1, A.2.1 and from the equality (4.2.5), it follows that there exist the integral

$$\int\left(N(x,y) - \frac{\partial}{\partial y}\int M(x,y)\circ dx\right)\circ dy$$
$$= \int N(x,y)\circ dy - \int\left(\frac{\partial}{\partial y}\int M(x,y)\circ dx\right)\circ dy$$

Therefore, we can find the map C_1 as solution of the differetial equation (4.2.13).

EXAMPLE 4.2.3. *Consider differential equation*

(4.2.16) $$(1\otimes 1 + 1\otimes y)\circ dx + (x\otimes 1 + 1\otimes 1)\circ dy = 0$$

Our goal is to find implicit function $u(x,y)$, which is solution of the differential equation (4.2.16). Differential equations

(4.2.17) $$\frac{\partial u}{\partial x} = 1\otimes 1 + 1\otimes y$$

(4.2.18) $$\frac{\partial u}{\partial y} = x\otimes 1 + 1\otimes 1$$

follow from the differential equation (4.2.16). Then

(4.2.19) $$u(x,y) = \int (1\otimes 1 + 1\otimes y)\circ dx = x + xy + C_1(y)$$

is the solution of the differential equation (4.2.17). The equality

(4.2.20) $$\frac{\partial u(x,y)}{\partial y} = x\otimes 1 + \frac{dC_1(y)}{dy} = x\otimes 1 + 1\otimes 1$$

follows from the equality (4.2.19) and from the equation (4.2.18). The differential equation

(4.2.21) $$\frac{dC_1(y)}{dy} = 1\otimes 1$$

follows from the equality (4.2.20).

(4.2.22) $$C_1(y) = y$$

is the solution of the differential equation (4.2.21). *From equalities* (4.2.19), (4.2.22), *it follows that implicit function*

$$x + xy + y = C$$

is the solution of the differential equation (4.2.16). □

The example 4.2.4 shows that equalities (4.2.4), (4.2.5) are essential for integralibility of differential equation (4.2.3).

EXAMPLE 4.2.4. *Consider the differential equation*

(4.2.23) $$(3x^2 \otimes 1 + 1 \otimes y) \circ dx + (x \otimes 1) \circ dy = 0$$

Here

$$M(x,y) = 3x^2 \otimes 1 + 1 \otimes y \qquad N(x,y) = x \otimes 1$$

Therefore

$$\frac{\partial M(x,y)}{\partial y} = 1 \otimes_x 1 \otimes_y 1$$

$$\frac{\partial N(x,y)}{\partial x} = 1 \otimes_x 1 \otimes_y 1$$

However, the differential equation (4.2.23) *is not integrable, because integral*

$$\int (3x^2 \otimes 1 + 1 \otimes y) \circ dx$$

does not exist. □

The example 4.2.5 shows that order of variables is essential in the equality (4.2.6).

EXAMPLE 4.2.5. *Consider the differential equation*

(4.2.24) $$(1 \otimes y) \circ dx + (1 \otimes x) \circ dy = 0$$

Here

$$M(x,y) = 1 \otimes y \qquad N(x,y) = 1 \otimes x$$

Therefore

$$\frac{\partial M(x,y)}{\partial y} = 1 \otimes_x 1 \otimes_y 1$$

$$\frac{\partial N(x,y)}{\partial x} = 1 \otimes_y 1 \otimes_x 1$$

Although the derivatives $\dfrac{\partial M(x,y)}{\partial y}$, $\dfrac{\partial N(x,y)}{\partial x}$ are represented by the same tensor, their action on variables x, y is different

$$\frac{\partial M(x,y)}{\partial y} \circ (dx, dy) = (1 \otimes_x 1 \otimes_y 1) \circ (dx, dy) = dx\, dy$$

$$\neq \frac{\partial N(x,y)}{\partial x} \circ (dy, dx) = (1 \otimes_y 1 \otimes_x 1) \circ (dy, dx) = dy\, dx$$

It is easy to see that the differential equation (4.2.24) is not integrable. $\square$

4.3. Linear Homogeneous Equation

THEOREM 4.3.1. *Let A be D-algebra. Let $a \in A \otimes A$. The map*

$$y = F \circ e^{a \circ x}$$

where $F \in A \otimes A$ satisfies to the equality

(4.3.1) $$F \circ 1 = C$$

is solution of the differential equation

(4.3.2) $$\frac{dy}{dx} - \frac{1}{2} a y - \frac{1}{2} y a = 0$$

with initial condition

$$x = 0 \quad y = C$$

PROOF. The equality

(4.3.3) $$\frac{dF \circ e^{a \circ x}}{dx} = F \circ \frac{a^{a \circ x}}{dx} = \frac{1}{2} F \circ (e^{a \circ x} a + a e^{a \circ x})$$

follows from theorems A.1.3, A.1.11. Therefore, the map

$$y = F \circ e^{a \circ x}$$

is solution of the differential equation (4.3.2). To find the tensor F, we set $x = 0$. The equality (4.3.1) follows from the equality $e^0 = 1$. $\square$

QUESTION 4.3.2. *Maps*

$$y = (C \otimes 1) \circ e^{a \circ x} = C e^{a \circ x}$$

$$y = (1 \otimes C) \circ e^{a \circ x} = e^{a \circ x} C$$

are solutions of the differential equation (4.3.2) *with initial condition*

$$x = 0 \quad y = C$$

In fact, this initial value problem has infinitely many solutions. It is important to understand the cause of this phenomenon. $\square$

THEOREM 4.3.3. *The differential equation* (4.3.2) *in quaternion algebra has form of system of differential equations*

(4.3.4)
$$\begin{cases} \dfrac{\partial y^0}{\partial x^0} = y^0 & \dfrac{\partial y^0}{\partial x^1} = -y^1 & \dfrac{\partial y^0}{\partial x^2} = -y^2 & \dfrac{\partial y^0}{\partial x^3} = -y^3 \\ \dfrac{\partial y^1}{\partial x^0} = y^1 & \dfrac{\partial y^1}{\partial x^1} = y^0 & \dfrac{\partial y^1}{\partial x^2} = 0 & \dfrac{\partial y^1}{\partial x^3} = 0 \\ \dfrac{\partial y^2}{\partial x^0} = y^2 & \dfrac{\partial y^2}{\partial x^1} = 0 & \dfrac{\partial y^2}{\partial x^2} = y^0 & \dfrac{\partial y^2}{\partial x^3} = 0 \\ \dfrac{\partial y^3}{\partial x^0} = y^3 & \dfrac{\partial y^3}{\partial x^1} = 0 & \dfrac{\partial y^3}{\partial x^2} = 0 & \dfrac{\partial y^3}{\partial x^3} = y^0 \end{cases}$$

with respect to the basis

$$e_0 = 1 \quad e_1 = i \quad e_2 = j \quad e_3 = k$$

PROOF. We can write differential equation (4.3.2) as follows

(4.3.5)
$$dy = \frac{1}{2} y\, dx + \frac{1}{2} dx\, y$$

4.3. Linear Homogeneous Equation

If we represent differentials dx, dy as vector-column, then, according to theorems [7]-5.1, [7]-5.2, the equation (4.3.5) gets following form

(4.3.6)
$$\begin{pmatrix} dy^0 \\ dy^1 \\ dy^2 \\ dy^3 \end{pmatrix} = \frac{1}{2} E_l(y) \begin{pmatrix} dx^0 \\ dx^1 \\ dx^2 \\ dx^3 \end{pmatrix} + \frac{1}{2} E_r(y) \begin{pmatrix} dx^0 \\ dx^1 \\ dx^2 \\ dx^3 \end{pmatrix}$$

$$= \frac{1}{2} \begin{pmatrix} y^0 & -y^1 & -y^2 & -y^3 \\ y^1 & y^0 & -y^3 & y^2 \\ y^2 & y^3 & y^0 & -y^1 \\ y^3 & -y^2 & y^1 & y^0 \end{pmatrix} \begin{pmatrix} dx^0 \\ dx^1 \\ dx^2 \\ dx^3 \end{pmatrix}$$

$$+ \frac{1}{2} \begin{pmatrix} y^0 & -y^1 & -y^2 & -y^3 \\ y^1 & y^0 & y^3 & -y^2 \\ y^2 & -y^3 & y^0 & y^1 \\ y^3 & y^2 & -y^1 & y^0 \end{pmatrix} \begin{pmatrix} dx^0 \\ dx^1 \\ dx^2 \\ dx^3 \end{pmatrix}$$

$$= \begin{pmatrix} y^0 & -y^1 & -y^2 & -y^3 \\ y^1 & y^0 & 0 & 0 \\ y^2 & 0 & y^0 & 0 \\ y^3 & 0 & 0 & y^0 \end{pmatrix} \begin{pmatrix} dx^0 \\ dx^1 \\ dx^2 \\ dx^3 \end{pmatrix}$$

Since the matrix of derivative has following form

$$\frac{dy}{dx} = \begin{pmatrix} \frac{\partial y^0}{\partial x^0} & \frac{\partial y^0}{\partial x^1} & \frac{\partial y^0}{\partial x^2} & \frac{\partial y^0}{\partial x^3} \\ \frac{\partial y^1}{\partial x^0} & \frac{\partial y^1}{\partial x^1} & \frac{\partial y^1}{\partial x^2} & \frac{\partial y^1}{\partial x^3} \\ \frac{\partial y^2}{\partial x^0} & \frac{\partial y^2}{\partial x^1} & \frac{\partial y^2}{\partial x^2} & \frac{\partial y^2}{\partial x^3} \\ \frac{\partial y^3}{\partial x^0} & \frac{\partial y^3}{\partial x^1} & \frac{\partial y^3}{\partial x^2} & \frac{\partial y^3}{\partial x^3} \end{pmatrix}$$

then the equality

$$(4.3.7) \quad \begin{pmatrix} \frac{\partial y^0}{\partial x^0} & \frac{\partial y^0}{\partial x^1} & \frac{\partial y^0}{\partial x^2} & \frac{\partial y^0}{\partial x^3} \\ \frac{\partial y^1}{\partial x^0} & \frac{\partial y^1}{\partial x^1} & \frac{\partial y^1}{\partial x^2} & \frac{\partial y^1}{\partial x^3} \\ \frac{\partial y^2}{\partial x^0} & \frac{\partial y^2}{\partial x^1} & \frac{\partial y^2}{\partial x^2} & \frac{\partial y^2}{\partial x^3} \\ \frac{\partial y^3}{\partial x^0} & \frac{\partial y^3}{\partial x^1} & \frac{\partial y^3}{\partial x^2} & \frac{\partial y^3}{\partial x^3} \end{pmatrix} = \begin{pmatrix} y^0 & -y^1 & -y^2 & -y^3 \\ y^1 & y^0 & 0 & 0 \\ y^2 & 0 & y^0 & 0 \\ y^3 & 0 & 0 & y^0 \end{pmatrix}$$

follows from (4.3.6). The system of differential equations (4.3.4) follows from the equality (4.3.7). □

THEOREM 4.3.4. *The system of differential equations* (4.3.4) *is not completely integrable.*

PROOF. Equalities

$$(4.3.8) \quad \frac{\partial^2 y^1}{\partial x^2 \partial x^1} = \frac{\partial}{\partial x^2} \frac{\partial y^1}{\partial x^1} = \frac{\partial y^0}{\partial x^2} = -y^2$$

$$(4.3.9) \quad \frac{\partial^2 y^1}{\partial x^1 \partial x^2} = \frac{\partial}{\partial x^1} \frac{\partial y^1}{\partial x^2} = \frac{\partial 0}{\partial x^2} = 0$$

follow from the system of differential equations (4.3.4). The statement

$$(4.3.10) \quad \frac{\partial^2 y^1}{\partial x^2 \partial x^1} \neq \frac{\partial^2 y^1}{\partial x^1 \partial x^2}$$

follows from equalities (4.3.8), (4.3.9). The theorem follows from the statement (4.3.10). □

The theorem 4.3.4 answers the question 4.3.2. However the new question arises. What is difference between the differential equation (4.3.2) and a differential equation considered in the theorem 3.5.5. Right side of the differential equation (3.5.7) depends only on x; this imposes a strong constrain on the existence of a solution. The differential equation (4.3.2) is an example of **linear homogeneous equation** of order 1. We can write linear homogeneous equation of order 1 as

$$(4.3.11) \quad \frac{dy}{dx} + (a_1 y a_2) \otimes a_3 + b_1 \otimes (b_2 y b_3) = 0$$

where $a_1, a_2, a_3, b_1, b_2, b_3 \in A$.

QUESTION 4.3.5. *Does there exist a constrain on A-numbers a_1, a_2, a_3, b_1, b_2, b_3 under the condition that the equation* (4.3.11) *has a solution? What is a solution of the equation* (4.3.11)? □

APPENDIX A

Summary of Statements

Let D be the complete commutative ring of characteristic 0.

A.1. Table of Derivatives

THEOREM A.1.1. *For any* $b \in A$
$$\frac{db}{dx} = 0 \otimes 0$$

PROOF. The theorem follows from the theorem [8]-B.1.1. □

THEOREM A.1.2.

(A.1.1) $$\frac{dx}{dx} \circ dx = dx \qquad \frac{dx}{dx} = 1 \otimes 1$$

PROOF. The theorem follows from the theorem [8]-B.1.2. □

THEOREM A.1.3. *For any* $F \in A \otimes A$,

(A.1.2) $$\begin{cases} \dfrac{dF \circ f(x)}{dx} = F \circ \dfrac{df(x)}{dx} \\ \dfrac{dF \circ f(x)}{dx} \circ dx = F \circ \left(\dfrac{df(x)}{dx} \circ dx \right) \end{cases}$$

PROOF. The theorem follows from the theorem [8]-B.1.3. □

THEOREM A.1.4. *Let*
$$f : A \to B$$
$$g : A \to B$$
be maps of Banach D-module A into associative Banach D-algebra A. Since there exist the derivatives $\dfrac{df(x)}{dx}, \dfrac{dg(x)}{dx}$, *then there exists the derivative* $\dfrac{d(f(x) + g(x))}{dx}$

(A.1.3) $$\frac{d(f(x) + g(x))}{dx} \circ dx = \frac{df(x)}{dx} \circ dx + \frac{dg(x)}{dx} \circ dx$$

(A.1.4) $$\frac{d(f(x) + g(x))}{dx} = \frac{df(x)}{dx} + \frac{dg(x)}{dx}$$

Since

(A.1.5) $$\frac{df(x)}{dx} = \frac{d_{s \cdot 0} f(x)}{dx} \otimes \frac{d_{s \cdot 1} f(x)}{dx}$$

$$\text{(A.1.6)} \qquad \frac{dg(x)}{dx} = \frac{d_{t \cdot 0} g(x)}{dx} \otimes \frac{d_{t \cdot 1} g(x)}{dx}$$

then

$$\text{(A.1.7)} \qquad \frac{d(f(x) + g(x))}{dx} = \frac{d_{s \cdot 0} f(x)}{dx} \otimes \frac{d_{s \cdot 1} f(x)}{dx} + \frac{d_{t \cdot 0} g(x)}{dx} \otimes \frac{d_{t \cdot 1} g(x)}{dx}$$

PROOF. The theorem follows from the theorem [8]-B.1.4. □

THEOREM A.1.5. *For any* $b, c \in A$

$$\text{(A.1.8)} \qquad \begin{cases} \dfrac{d\,bxc}{dx} = b \otimes c & \dfrac{d\,bxc}{dx} \circ dx = b\,dx\,c \\[2mm] \dfrac{d_{1 \cdot 0} bxc}{dx} = b & \dfrac{d_{1 \cdot 1} bxc}{dx} = c \end{cases}$$

PROOF. Corollary of theorems A.1.2, A.1.3, when $f(x) = x$. □

THEOREM A.1.6. *Let f be linear map*

$$f \circ x = (a_{s \cdot 0} \otimes a_{s \cdot 1}) \circ x = a_{s \cdot 0}\, x\, a_{s \cdot 1}$$

Then

$$\frac{\partial f \circ x}{\partial x} = f$$

$$\frac{\partial f \circ x}{\partial x} \circ dx = f \circ dx$$

PROOF. Corollary of theorems A.1.4, A.1.5, [8]-3.3.13. □

COROLLARY A.1.7. *For any* $b \in A$

$$\begin{cases} \dfrac{d(xb - bx)}{dx} = 1 \otimes b - b \otimes 1 \\[2mm] \dfrac{d(xb - bx)}{dx} \circ dx = dx\,b - b\,dx \\[2mm] \dfrac{d_{1 \cdot 0}(xb - bx)}{dx} = 1 & \dfrac{d_{1 \cdot 1}(xb - bx)}{dx} = b \\[2mm] \dfrac{d_{2 \cdot 0}(xb - bx)}{dx} = -b & \dfrac{d_{2 \cdot 1}(xb - bx)}{dx} = 1 \end{cases}$$

□

A.1. Table of Derivatives

THEOREM A.1.8. *Let D be the complete commutative ring of characteristic 0. Let A be associative Banach D-algebra. Then*[A.1]

(A.1.9) $$\frac{dx^2}{dx} = x \otimes 1 + 1 \otimes x$$

(A.1.10) $$dx^2 = x\,dx + dx\,x$$

(A.1.11) $$\begin{cases} \dfrac{d_{1 \cdot 0} x^2}{dx} = x & \dfrac{d_{1 \cdot 1} x^2}{dx} = 1 \\ \dfrac{d_{2 \cdot 0} x^2}{dx} = 1 & \dfrac{d_{2 \cdot 1} x^2}{dx} = x \end{cases}$$

PROOF. The theorem follows from the theorem [8]-B.1.15. □

THEOREM A.1.9. *Let D be the complete commutative ring of characteristic 0. Let A be associative Banach D-algebra. Then*

(A.1.12) $$\frac{dx^3}{dx} = x^2 \otimes 1 + x \otimes x + 1 \otimes x^2$$

(A.1.13) $$dx^3 = x^2\,dx + x\,dx\,x + dx\,x^2$$

PROOF. The theorem follows from the theorem [8]-B.1.17. □

THEOREM A.1.10.

(A.1.14) $$\frac{de^x}{dx} = \frac{1}{2}(e^x \otimes 1 + 1 \otimes e^x)$$

PROOF. The theorem follows from the theorem [8]-5.2.7. □

THEOREM A.1.11. *Let* $a \in A \otimes A$.

(A.1.15) $$\frac{de^{a \circ x}}{dx} = \frac{1}{2}(e^{a \circ x} a + a e^{a \circ x})$$

PROOF. The equality
$$\frac{de^{a \circ x}}{dx} = \frac{de^{a \circ x}}{da \circ x} \circ \frac{da \circ x}{dx} = \frac{1}{2}(e^{a \circ x} \otimes 1 + 1 \otimes e^{a \circ x}) \circ a = \frac{1}{2}(e^{a \circ x} a + a e^{a \circ x})$$
follows from theorems A.1.10, [8]-3.3.23. □

[A.1] The statement of the theorem is similar to example VIII, [20], p. 451. If product is commutative, then the equality (A.1.9) gets form
$$dx^2 \circ dx = 2x\,dx$$
$$\frac{dx^2}{dx} = 2x$$

THEOREM A.1.12.

(A.1.16) $$\frac{d\sinh x}{dx} = \frac{1}{2}(\cosh x \otimes 1 + 1 \otimes \cosh x)$$

(A.1.17) $$\frac{d\cosh x}{dx} = \frac{1}{2}(\sinh x \otimes 1 + 1 \otimes \sinh x)$$

PROOF. The theorem follows from the theorem [8]-5.3.2. □

THEOREM A.1.13.

(A.1.18) $$\frac{d\sin x}{dx} = \frac{1}{2}(\cos x \otimes 1 + 1 \otimes \cos x)$$

(A.1.19) $$\frac{d\cos x}{dx} = -\frac{1}{2}(\sin x \otimes 1 + 1 \otimes \sin x)$$

PROOF. The theorem follows from the theorem [8]-5.4.2. □

THEOREM A.1.14. *Let A be Banach D-module. Let B be Banach D-algebra. Let f, g be differentiable maps*

$$f : A \to B \quad g : A \to B$$

The derivative satisfies to relationship

$$\frac{df(x)g(x)}{dx} \circ dx = \left(\frac{df(x)}{dx} \circ dx\right) g(x) + f(x) \left(\frac{dg(x)}{dx} \circ dx\right)$$

(A.1.20) $$\frac{df(x)g(x)}{dx} = \frac{df(x)}{dx} g(x) + f(x) \frac{dg(x)}{dx}$$

PROOF. The theorem follows from the theorem [8]-3.3.17. □

THEOREM A.1.15. *Let A be Banach D-module. Let B, C be Banach D-algebras. Let f, g be differentiable maps*

$$f : A \to B \quad g : A \to C$$

The derivative satisfies to relationship

$$\frac{df(x) \otimes g(x)}{dx} \circ a = \left(\frac{df(x)}{dx} \circ a\right) \otimes g(x) + f(x) \otimes \left(\frac{dg(x)}{dx} \circ a\right)$$

$$\frac{df(x) \otimes g(x)}{dx} = \frac{df(x)}{dx} \otimes g(x) + f(x) \otimes \frac{dg(x)}{dx}$$

PROOF. The theorem follows from the theorem [8]-3.3.20. □

A.2. Table of Integrals

THEOREM A.2.1. *Let the map*
$$f : A \to B$$
be differentiable map. Then

(A.2.1) $$\int \frac{df(x)}{dx} \circ dx = f(x) + C$$

PROOF. The theorem follows from the definition 3.1.1. □

THEOREM A.2.2. *Let the map*
$$f : A \to \mathcal{L}(D; A; B)$$
be integrable map. Then

(A.2.2) $$\frac{d}{dx} \int f(x) \circ dx = f(x)$$

PROOF. The theorem follows from the definition 3.1.1. □

THEOREM A.2.3.

(A.2.3) $$\int (0 \otimes 0) \circ dx = C$$

PROOF. The theorem follows from the theorem A.1.1. □

THEOREM A.2.4.

(A.2.4) $$\int (f(x) + g(x)) \circ dx = \int f(x) \circ dx + \int g(x) \circ dx$$

PROOF. The equlity (A.2.4) follows from theorems A.1.4, A.2.1. □

THEOREM A.2.5.

(A.2.5) $$\int (f_{s \cdot 0} \otimes f_{s \cdot 1}) \circ dx = (f_{s \cdot 0} \otimes f_{s \cdot 1}) \circ x + C$$

(A.2.6) $$\int f_{s \cdot 0} \, dx \, f_{s \cdot 1} = f_{s \cdot 0} \, x \, f_{s \cdot 1} + C$$

$$f_{s \cdot 0} \in A \quad f_{s \cdot 1} \in A$$

PROOF. The theorem follows from the theorem [8]-B.2.4. □

THEOREM A.2.6.

(A.2.7) $$\int (1 \otimes x + x \otimes 1) \circ dx = x^2 + C$$

(A.2.8) $$\int dx \, x + x \, dx = x^2 + C$$

THEOREM A.2.7.

(A.2.9) $$\int (1 \otimes x^2 + x \otimes x + x^2 \otimes 1) \circ dx = x^3 + C$$

(A.2.10) $$\int dx\, x^2 + x\, dx\, x + x^2\, dx = x^3 + C$$

PROOF. The theorem follows from the theorem [8]-5.1.3. □

THEOREM A.2.8.

(A.2.11) $$\int (e^x \otimes 1 + 1 \otimes e^x) \circ dx = 2e^x + C$$

(A.2.12) $$\int e^x\, dx + dx\, e^x = 2e^x + C$$

PROOF. The theorem follows from the theorem [8]-B.2.7. □

THEOREM A.2.9.

(A.2.13) $$\int (\sinh x \otimes 1 + 1 \otimes \sinh x) \circ dx = 2\cosh x + C$$

(A.2.14) $$\int \sinh x\, dx + dx\, \sinh x = 2\cosh x + C$$

(A.2.15) $$\int (\cosh x \otimes 1 + 1 \otimes \cosh x) \circ dx = 2\sinh x + C$$

(A.2.16) $$\int \cosh x\, dx + dx\, \cosh x = 2\sinh x + C$$

PROOF. The theorem follows from the theorem [8]-B.2.8. □

THEOREM A.2.10.

(A.2.17) $$\int (\sin x \otimes 1 + 1 \otimes \sin x) \circ dx = -2\cos x + C$$

(A.2.18) $$\int \sin x\, dx + dx\, \sin x = -2\cos x + C$$

(A.2.19) $$\int (\cos x \otimes 1 + 1 \otimes \cos x) \circ dx = 2\sin x + C$$

(A.2.20) $$\int \cos x\, dx + dx\, \cos x = 2\sin x + C$$

PROOF. The theorem follows from the theorem [8]-B.2.9. □

References

[1] Serge Lang, Algebra, Springer, 2002
[2] James Stewart, Calculus,
Cengage Learning, 2012, ISBN: 978-0-538-49781-7
[3] William E. Boyce, Richard C. DiPrima, Elementary Differential Equations and Boundary Value Problems,
John Wiley & Sons, Inc., 2009, ISBN 978-0-470-38334-6
[4] Lev Elsgolts, Differential Equations and the Calculus of Variations,
Translated from the Russian by George Yankovsky,
MIR Publishers, Moscow, 1977
[5] Aleks Kleyn, Introduction into Calculus over Division Ring,
eprint arXiv:0812.4763 (2010)
[6] Aleks Kleyn, Linear Maps of Free Algebra,
eprint arXiv:1003.1544 (2010)
[7] Aleks Kleyn, Linear Maps of Quaternion Algebra,
eprint arXiv:1107.1139 (2011)
[8] Aleks Kleyn.
Single Variable Calculus: Noncomutative Banach Algebra.
CreateSpace Independent Publishing Platform, 2014;
ISBN-13: 978-1497563810
[9] Aleks Kleyn.
Linear Algebra over Division Ring: Vector Space.
CreateSpace Independent Publishing Platform, 2014;
ISBN-13: 978-1499324006
[10] Aleks Kleyn.
Representation of Universal Algebra: Polymorphism.
CreateSpace Independent Publishing Platform, 2015;
ISBN-13: 978-1511460194
[11] John C. Baez, The Octonions,
eprint arXiv:math.RA/0105155 (2002)
[12] Paul M. Cohn, Universal Algebra, Springer, 1981
[13] Paul M. Cohn, Algebra, Volume 1, John Wiley & Sons, 1982
[14] Eisenhart, Continuous Groups of Transformations, Dover Publications, New York, 1961
[15] Postnikov M. M., Geometry IV: Differential geometry, Moscow, Nauka, 1983

[16] Fikhtengolts G. M., Differential and Integral Calculus Course, volume 1, Moscow, Nauka, 1969

[17] Fikhtengolts G. M., Differential and Integral Calculus Course, volume 2, Moscow, Nauka, 1969

[18] Alekseyevskii D. V., Vinogradov A. M., Lychagin V. V., Basic Concepts of Differential Geometry
VINITI Summary 28
Moscow. VINITI, 1988

[19] Richard D. Schafer, An Introduction to Nonassociative Algebras, Dover Publications, Inc., New York, 1995

[20] Sir William Rowan Hamilton, Elements of Quaternions, Volume I, Longmans, Green, and Co., London, New York, and Bombay, 1899

Index

A-number 6
A-representation in Ω-algebra 6
algebra over ring 10
arity 5
associative D-algebra 11
associative law 8
associator of D-algebra 11
norm of quaternion 16

Banach D-algebra 13

Cartesian power 5
center of D-algebra A 11
commutative D-algebra 11
commutator of D-algebra 11
continuous map 13
coordinates 8
coproduct of objects in category 16

D-algebra 10
D-module 8
derivative of map 14
derivative of order n 15
derivative of second order 15
differentiable map 14
differential equation with separated variables 49
differential of map 14
differential separable equation 48
direct sum 17, 18
distributive law 8

effective representation 6
endomorphism 6
exact differential equation 51

free algebra over ring 10

homomorphism 6

indefinite integral 27
integrable differential equation 27
integrable map 27

linear homogeneous equation 58
linear map 8, 12

map is compatible with operation 6
matrix of linear map 9
method of successive differentiation 36
module over ring 8
morphism of representation f 7
morphism of representations from f into g 7
morphism of representations of Ω_1-algebra in Ω_2-algebra 7

n-ary operation on set 5
norm of map 13
nucleus of D-algebra A 11

operation on set 5
operator domain 5

partial derivative 24
partial derivative of second order 26
partial linear map 21
polylinear map 9, 12

quaternion algebra 15

reduced morphism of representations 7
representation of Ω_1-algebra A in Ω_2-algebra M 6

structural constants 11

tensor product 10

unitarity law 8
universal algebra 5

vector 8

Ω-algebra 5

Special Symbols and Notations

$A \oplus B$ direct sum 17, 18

(a, b, c) associator of D-algebra 11
$[a, b]$ commutator of D-algebra 11
A_Ω Ω-algebra 5
B^A Cartesian power 5
$B_1 \coprod ... \coprod B_n$ coproduct in category 17

C_{ij}^k structural constants 11

$\dfrac{df(x)}{dx}$ derivative of map f 14
$d_x f(x)$ derivative of map f 14
$\dfrac{d^n f(x)}{dx^n}$ derivative of order n 15
$d_{x^n}^n f(x)$ derivative of order n 15
$\dfrac{d^2 f(x)}{dx^2}$ derivative of second order 15
$d_{x^2}^2 f(x)$ derivative of second order 15
dx differential of independent variable 14
df differential of map f 14
$\dfrac{\partial f^i}{\partial x^j}$ partial derivative 24
$\dfrac{\partial^2 f}{\partial x^i \partial x^j}$ partial derivative of second order 26

$\mathrm{End}(\Omega; A)$ set of endomorphisms 6

$\|f\|$ norm of map 13

$\displaystyle\int g(x) \circ dx$ indefinite integral 27

H quaternion algebra 15

$\mathcal{L}(D; A_1 \to A_2)$ set of linear maps 8, 12
$\mathcal{L}(D; A_1 \times ... \times A_n \to S)$ set of polylinear maps 12
$\mathcal{L}(D; A^n \to S)$ set of n-linear maps 12

$N(A)$ nucleus of D-algebra A 11

$A_1 \otimes ... \otimes A_n$ tensor product 10

$Z(A)$ center of D-algebra A 11

Ω operator domain 5
$\Omega(n)$ set of n-ary operators 5

$\displaystyle\coprod_{i \in I} B_i$ coproduct in category 16

$\displaystyle\coprod_{i=1}^{n} B_i$ coproduct in category 17

www.ingramcontent.com/pod-product-compliance
Lightning Source LLC
Chambersburg PA
CBHW051202220526
45473CB00003B/878